LIGHT

REPLACING THREE CENTURIES OF MISCONCEPTIONS

Peter Bros

1996

INFORMATION
PHONE 1-703-971-9141
FAX 1-703-971-1628
info@copernican-series.com
WEB SEARCH: "copernican series"

PUBLISHED BY FINANCIAL BOOK PARTNERS

PRINTED IN THE UNITED STATES OF AMERICA

ISBN 0-9627769-6-3

Designed by Lemieux Creative Enterprises

LIGHT

REPLACING THREE CENTURIES OF MISCONCEPTIONS

THE COPERNICAN SERIES

AT THE GATES OF THE CITADEL

THE SUBJUGATION OF MODERN SCIENCE

How the scientific method generates facts, conceptually blinding us to reality

THE COOLING CONTINUUM

THE RISE AND FALL OF SPECIES ON EARTH

How life forms and evolves and why it becomes extinct

ATOMS, STARS AND MINDS

SYNTHESIZING AN ELEMENTARY PARTICLE THAT COMPREHENDS ITSELF

How matter forms and how it moves in space

THE MODEL MIND

HOW THE MIND MOVES MATTER

How the mind is constructed and how it forms, stores and recalls images

HOW THE WEATHER REALLY WORKS!

How air masses transport heat, and what transports the air masses

LIGHT

REPLACING THREE CENTURIES OF MISCONCEPTIONS

Unifying light, heat, electricity and magnetism with a single concept

1. What Is Light?

Empirical science sits down and looks at something and then goes through a process of monkey see, monkey say. Monkeys have a similar process. They see something done, form a picture in their mind of themselves doing it, and then do it. No intelligence intervenes between the monkey see and the monkey do. Empirical science sees something, forms a mental picture of what it sees, and instead of doing something, makes up something to say about what it sees.

There is no application of intelligence between what empirical science sees and what it says.

As a result, we see something drop and we say it drops because it is a property for it to drop (it drops because it drops), we see the sun move across the sky and we say the sun travels around the Earth and, when we realize that it is the Earth that is rotating rather than the sun that is orbiting, we say the Earth rotates as a result of its angular momentum (it rotates because it rotates), we see the planets orbit and say they condensed out of a swirling mass of gas (they orbit because they orbit), we see two ends of a magnet attract and two ends repel and we say that the ends are opposite poles so that like poles repel while opposite poles attract (the ends attract and repel because of polarity, because they have opposite ends) and when we see electricity generated in one place act over a circuit in another place, we say that electricity is caused by a charged particle called an electron (electricity is a moving charge).

And, when we see sunlight break up into the colors of a rainbow, like monkeys who might finger the television screen in an attempt to get at the little figures inside the box, we say that all of the colors of the rainbow are contained in the sunlight.

Where else could they have come from?

And, when we divide light and then intermingle it and we see alternating lines of light and no light, we mindlessly look around for a picture that might explain it and say that light is like water waves.

We conduct our exploration of reality like monkeys, imitating what we see in reality with words rather than acts, and as a result, we have produced a science that is suited to monkeys, a science based on monkey see, monkey say.

Because empirical science merely names, classifies, gives things designations, defines and categorizes what it sees in physical reality, it does not deal with comprehension. It is not concerned with the meaning of facts in reality and how those facts operate to produce the physical reality that we measure.

Yet facts are meaningless without comprehension.

The most important aspect of physical reality, what makes things move with respect to other things in physical reality, is not directly measurable. Because empirical science has not been able to come up with meaningful explanations for the movement of matter in physical reality, what it is in physical reality that makes things fall, what it is in physical reality that makes the planets and moons rotate on their axes and what makes moons orbit planets and planets orbit the sun, it has concluded that simply measuring what exists is sufficient. Meaning has no validity unless it has been used to predict another fact, in which case it ceases to be meaning and becomes a fact.

The scientific method, which is useful only in verifying whether, in fact, facts exist in reality, has been used as a way to make us believe that we know what we cannot know and as a result, generates facts rather than explains them. We make up concepts about the facts we see, matter and the movement of matter in space, concepts that have nothing to do with those facts, when the matter came into existence, how far it extends in space, and when it is going to come to an end and then look for facts those concepts predict. Because they are made-up concepts, it isn't too difficult to find made-up facts to confirm them, or, if nothing is predictable, make up a connection between exiting fact and the made-up concepts that will allow us to believe the concept actually predicts the fact.

What this process does is generate facts which are really concepts, because our concepts can never be made real other than in our own minds. Concepts are what we use to arrange the facts we measure in the universe, and they are only as good as the facts they explain. Statements such as an object will move in a straight line so long as another force doesn't act on it to change that motion are simply made-up concepts that have nothing to do with facts in reality. In reality, objects do not move in straight lines, they come to rest with respect to the forces acting upon them. In like manner, statements such as planets rotate and orbit because of angular momentum, gravity is a property of matter, the solar system condensed out of a swirling mass of gas, electricity is a moving charge, magnets operate on the basis of polarity, light is analogous to water waves, all of these statements are just meaningless gibberish designed to cover up the fact that we don't actually know what we are talking about, that we have no comprehension of the facts that we are using the gibberish to explain.

When we don't have in our arsenal any way to deal with comprehension, when we have a procedure that turns our fantasies into facts, then we have no way of separating fact from our fantasies. When we attempt to come up with a consistent picture of physical reality, which should be our goal in the effort to come to grips with the reality in which we find ourselves, we end up dealing with a lot of facts that simply do not exist in reality.

Because we have made up so many facts, facts which are often contradictory and mostly confusing, we think we have an impossible number of facts with which to deal. However, if we deal with the facts in reality rather than the scientific facts our own reality has manufactured, we find that the facts that we know are facts in physical reality are really quite limited.

We know that there is a force that causes objects to move toward the surface of the Earth. We know that there are different types of matter because it takes different levels of force to move different objects against that force. However, we also know that objects that move freely within the force all move at the same rate, and we can measure this rate, and therefore the force, as being inversely proportional to the square of the distance from the source toward which the objects are moving.

That's about it for the surface of the Earth.

In space, we know that the planets orbit the sun and that moons orbit planets, and because we know this, we know, or should know, that there is a current force moving these objects in space.

We also know that the planets generally orbit the sun in the same direction, as do the moons the planets, and the planets and moons generally all rotate on their axes in the same direction as they orbit. The sun also rotates in this same direction, and the planets orbit and rotate in the equatorial plane produced by the sun's rotation.

That's about it for the solar system.

Although we can see groups of stars called galaxies that appear to be flattening out in an open solar system type configuration, we have no way of knowing how far these objects are, how big they are, or even in what direction they are rotating and orbiting.

Further, we know that there must be some sort of

attractive force that holds matter together because we live in a world of solid matter. And we know that there must be a force that allows matter to come apart because we see matter coming apart in space, that's how we know there is space because of the light the sun and the other stars produce, and we can build fires on the surface of the Earth and create technology that allows us to produce similar light.

We know that when matter comes apart, it produces light.

We know quite a bit about light, which is only right as light is what allows us to know about everything else.

We know, or should know, that light and heat are the same.

We know, or should know, that light must be made up of the same thing that gives it off.

We know and can measure that light is given off on the surface of matter undergoing an energetic process, burning or combustion.

On a more direct level, we know that when light is given off by matter, it expands away from that matter in a sphere. Because of this, we know, and can measure, that light diminishes as it moves away from the matter that is producing it. We know exactly how it diminishes, inversely proportional to the square of the distance it has moved away from that surface, the exact same measurement we have for the force that causes objects to fall.

We know that when light diminishes inversely with the square of the distance over which it expands, it does so uniformly.

We know a lot of other things about light.

We know that it is produced in a process that gives it frequency, a process in which the surface of matter expands and contracts as the light is given off.

We know that light comes in different frequencies which the eye senses as color, just as the skin senses different degrees of frequency in the form of heat. If we pass light through conductors fabricated into prisms, we can see it emerge in different frequencies, the spectrum of the rainbow.

More telling, we know that light travels at a certain speed before it enters the prism, slows down while it is moving through the prism, and regains its speed when it re-emerges from the prism.

This means that a property of what makes light up is its speed, that it is moving at the speed of light when it is at rest. This tells us something else about what makes up light because what makes up light also makes up matter. It tells us that what is holding matter together must also be a property of the same thing that makes up light so that when matter exists, the property that holds matter together has overcome the property that makes light move and the potential for movement is in whatever is making up the matter. Light is the product of releasing matter from the property that is holding it together so that it regains its at rest speed.

We know something else about light, something very important. We know that if we pass light through two slits and allow it to intermingle, that it recombines, falling on a collection screen in areas of recombined bright light and no light. Knowing this allows us to know that there is something about expanding flows of light that allows them to recombine as part of the process of expansion, and thus to diminish uniformly as they expand.

We know that there are frequencies of light that occur below the frequencies our eyes can sense as color. We can use electricity and magnets or electromagnets to create an oscillation that mimics the process in which matter produces light, producing radio waves which have frequencies below what the eye can see.

And while it is generally believed that Franklin proved that lightning was electricity and not fire, forever precluding us from thinking about light and electricity in the same mental picture, we know that when we get down to the lowest frequency possible, no frequency, the flow of electricity in a conductor, that flow creates a secondary inductive flow that orbits the flow at right angles. We know of this inductive flow because it can pull two conductors together or push them apart depending on the direction of the flows the conductors contain, just as magnetic lines of force can pull magnets together or push them apart.

Other things we know about light: We know that it bounces off matter, such as the light bouncing off of a mirror. We know that bouncing light off surfaces or passing it through certain crystals polarizes it, which is to say organizes it in certain ways.

We know that the lower a frequency, the greater it can penetrate matter, while high frequencies can disorganize the surface of the same matter. A match flame will burn our finger while radio waves will move right through it. Blue light penetrates matter, releasing electrons from the atoms that make up the matter itself while red, yellow or white light will break down on the surface, producing large quantities of disorganized electrons, and, if the frequency is high enough, disorganizing the atoms that make up the surface of the matter.

We know that reflecting a match flame off of matter will lower its frequency so that the light will no longer have a high enough frequency to destroy the surface of the matter. In addition, we know that the positioning of the atoms on the surface of matter determines the frequency of the light that results after it bounces off the matter, so that light that bounces off of the surface of matter changes frequency in one manner (the printing colors) while the light conducted by matter changes frequency in another manner (the colors that emerge from a prism).

While it is generally felt that light has no mass, a term made up to explain why there is no correlation between the size of the objects in the universe and the designated cause of gravity, the matter that makes up those objects, light has a demonstrable physical presence. The strength of light is regularly measured by how much force it exerts on a mirrored paddle in a vacuum.

All of what we know about physical reality is a result of light, and more than half of what we know about physical reality is directly about light. Yet we actually know nothing about how physical reality operates because we have accepted monkey see, monkey say answers for light. We say light is analogous to water waves even though water waves require a medium and light clearly travels without a medium. We say that white light contains all of the colors we see broken out by a prism, and that color is the result of matter absorbing all colors but the color we see, even though such an idea is absurd, requiring as it would matter to absorb more frequencies than it reflects. And we say that what makes up matter is not what makes up light because we have concluded that matter has mass that produces gravity and weight, and in our total confusion between the two effects, one which causes objects to drop, the other which is the result of the force required to move objects against the force that causes objects to drop, we require matter that is giving off light to change weight as it does so to prove that what makes up matter makes up light.

We attach unexamined ideas to physical facts that we are capable of measuring on a repeatable basis, and recall them mindlessly whenever those facts are measured. Because we see the world with light, we create a world out of our misconception of what light is. Just as we reverse the reality of the large, the macrocosm, by believing that objects move in a straight line unless a force acts on them to change that motion, the most fundamental error anyone could make about the physical universe, the misconception of light as water waves has literally warped our conception of the small, the microcosm we construct to explain the physical universe.

This is expressed in the prevailing view that Newtonianism, combining the fallacy of perpetual motion with the occult concept that a force that actually does mechanical work can be a nonmechanical property of matter, works only in the realm of physical reality, and that other, even more esoteric laws take over in the quantum realm we construct in our minds, a reality that, while making up physical reality, doesn't operate like physical reality.

The microcosmic misconception of light as water waves is based as firmly in Newtonianism, however, as the mass/gravity fallacy because Newton was the source of the misconception about the nature of white light. He concluded that the colors that came out of the prism were merely hidden in the white light rather than the colors being a product of the frequency of the light itself becoming progressively lower as the light traveled through progressively longer stretches of the material that made up the prism. This misconception allowed Young, as a result of his famous two-slit experiment which demonstrated the alternating areas of light and no light, the interference patterns formed when light is divided, recombined and allowed to collect on a screen, to further misconceive light as analogous to water waves, complete with crests and troughs which canceled each other out to provide the calm areas of no light on the collection screen.

Water waves, of course, have no independent existence, being merely a distortion of a medium through which competing forces are expressing themselves. However, in the monkey see, monkey say realm of empirical science, once the analogy was made, whenever we saw light, we visualized it as traveling as water waves even though the visualization reduced what is a three dimensional reality to a two dimensional surface distortion in our minds.

Because empirical science does not recognize the fact that comprehension rests on the formulation of concepts, but rather chooses to delude itself that some hypotheses are facts because they predicted facts which were later found to be in accord with the hypotheses, the misconception of light as being made up of a stack of wavelengths which travel like waves across the surface of water was used, not as a hypothesis in search of other facts, but as an unexamined fact, a fact assumed to exist, and therefore capable of appearing silently, invisibly, in hypotheses formulated to examine other aspects of physical reality.

When we then bent light to our use as a measuring device, we assumed that we knew everything there was to know about light. When our measurements failed to accord with the reality that we could see and otherwise measure, we didn't think to check our understanding of the measuring device, our understanding of light. We simply conformed the universe to our misunderstanding.

This is what happened when Michelson and Morley set out to measure the absolute speed of the Earth and the direction that it was traveling through space. With the analogy of light to water waves, all they had to do was measure the motion of the Earth against the movement of the light waves in the medium through which the Earth had to be moving. The feature of light used to measure the relative motion were interference patterns, the lines of light and no light that appeared on a collection screen when light was split and intermingled in Young's two-slit experiment.

This is empirical mindlessness in its highest expression!

The only thing we really know is that allowing split light to intermingle results in the light hitting the collection screen in bands of light and no light. What is actually happening between the time the light is split and the time it hits the collection screen is not directly measurable. The only thing that is measurable is that the intermingled light has recombined in such a way that, when the light falls on the collection screen, it is concentrated in areas of light between areas of no light.

Analogizing light to water waves and then creating a medium through which it has to travel doesn't tell us anything about what is actually happening to the light between the time it is split and the time it hits the collection screen. To understand what is actually happening, we need to have an idea about what light is, how it is produced, what it is made out of, how it travels and what happens when it hits something.

We have to have a picture of the structure of light.

Seeing light and saying water waves, however, allows empirical science to conclude that anything it wants to comprehend about light it can find out by measuring water waves. As a result, we end up measuring water waves and leave unmeasured the realities of light, the most obvious measurement being that it exists in three-dimensional space rather than merely being a three-dimensional disturbance in two-dimensional space.

If we had stopped Michelson or Morley when they were setting up the equipment for their memorable attempt to measure the absolute motion of the Earth and said, how is it that light is produced, how does it move through space, what makes it diminish uniformly as it does so, what is actually happening to its structure as it bounces off the mirrors of your device, their answer would have been, these questions are irrelevant to the structure and logic of the experiment. We are merely using the aether, the medium through which light travels, to measure the motion of the Earth.

But, we would reply, just saying there is an aether assumes that you know the answers to all of the unanswered questions. Assuming away answers only hides the questions.

We are not trying to play God, here, comes the empirical chorus. We are just attempting to determine a physical reality, the absolute motion of the Earth in space. Science does not concern itself with philosophical questions, it concerns itself with physical facts.

But, we insist, you're using philosophical answers to assume away basic questions of fact just as Newton used God and we use swirling masses of gas to assume away basic questions of motion in the solar system.

The use by empirical science of light as a measuring device without knowing what the nature of light is, is a mind-dead operation, one that would not be tolerated in even the most primitive of societies because it would raise the suspicion of trickery and thievery, and when discovered could lead to rapid retribution. The King kept the master weights and measures under lock and key so that his subjects would always have a reference point for their commercial transactions, and violating the weights and measures, in effect cheating our fellow out of his due, would lead to royal penalties if we hadn't already been lynched.

It is even more reprehensible to use light as a measuring device without knowing what light is because the loss is not of an ounce of grain or a fleck of gold, but rather the loss is of understanding, the abysmal ignorance that results from thinking we know something when in fact we know nothing.

Ignorance may well be bliss, but it is not conducive to survival.

Newtonianism is also responsible for the incredible misconception that light is not made up of what gives it off. While the misconceptions that white light is a mixture of all of the colors and that light travels in waves are merely mindless explanations that preclude alternate explanations, the idea that light is not made up of what gives it off is simply a negative that precludes any attempt to analyze what light really is.

The ignorance created by all of the great empirical misconceptions, objects travel in a straight line unless a force acts to change that motion, mass is proportional to and a property of mass, the solar system formed out of a swirling mass of gas, angular momentum is conserved in closed systems, electricity is a moving charge, attraction and repulsion are produced by polarity and light travels as water waves, untestable hypotheses all, is as nothing when compared to the ignorance created by the simple edict that light is not made up of what gives it off.

This edict arose out of arguments centering around whether light was actually something that moved from one place to another or whether it was the disturbance of a medium. The question paralleled the argument about whether light was a particle or a wave. If light was a particle, then it was made up of what produced it and moved from cause to effect. If light was a wave, then it had no independent existence other than as a disturbance of the medium through which it moved.

Lavoisier used the concept of combustion as the chemical combination of oxygen and fuel atoms to kill caloric theory. At the same time, he attempted to eliminate all concepts of light along with heat as something that moved from one place to another. To do this, he claimed that for light to exist between cause and effect it must be matter. If it was matter, it must have mass. For matter to have mass, it must have weight, confusing mass, the matter created by applying an inverse square law to the motion of matter in space with weight, the force it takes to keep an object from moving in a field of attraction. Lavoisier then purported to prove that the production of light by burning matter did not produce a change in the weight of the matter. Light therefore could have no mass, was insubstantial, was, in fact, an imponderable and therefore did not exist between cause and effect.

Whatever light was, then, it wasn't the same thing that made up matter because all matter had to have mass.

A child could conclude, as Newton did, that light is made up of a rainbow, and an observant child could come up with the analogy of light to water waves although a really observant child would discard both almost immediately.

It takes an educated child to come up with the conclusion that light is not made out of the same thing that gives it off. This sort of empirical nonsense has to be taught. It cannot be learned from observation.

All of this doesn't get us any closer to an explanation of

what light is, but before we can see light for what it is, we have to rid ourselves of our misconceptions about what we think light is.

The following three chapters will therefore deal with each of these misconceptions, white light as made up of all colors, light as water waves, and light as not being made up of the same thing that gives it off, in detail. The chapter following will then deal with a basic misconception about the matter that produces light, the notion that matter embodies polarity.

We have to eliminate our misconceptions about light and its source if we want to understand what light is.

2. The Eye Didn't Evolve To See Rainbows

Newton experimented with the light a prism produced. Allowing white light to pass through the prism produced colored light, with the light lined up from the top of the prism, red on top, then yellow and green, and then blue on the bottom. By isolating the light, Newton found that the prism only changed white light and that recombining the colored light on a screen which reflected all colors without change produced white light once again. Reflecting red and green light off such a white standard screen produced yellow light and adding blue light to the yellow light produced white light.

This is the type of empirical observation that can be used as a substitute for comprehension, and Newton concluded that the white light contained all colors. What the prism produced from white light empirical science could recombine back into white light and thus white light was a combination of all colors.

Next observing the order in which the light emerged from the prism, with the blue light making the greatest apparent angle as it emerged, and the red making the least, Newton concluded that the gravity of the corpuscles in the base of the prism, where most of the corpuscles of matter were located, was causing the corpuscles of light to speed up and deflect in that direction.

Those opposed to Newton's corpuscle theory of light,

the people that thought light traveled in a wave, concluded that Newton had it wrong with respect to the light speeding up. Applying Newton's successful inverse square law of gravity to the velocity of light, wave advocates concluded that the velocity of light was inversely proportional to its refractive index, the measurable amount that light bent when it passed from one medium to another. Light was therefore slowing down instead of speeding up when it entered the prism. The blue light was slowing down the most, with the green and yellow slowing down progressively less while the least refracted, red, was slowing down the least.

The wave advocates did not think to question Newton's conclusion that white light contained all colors and no one has ever bothered to question the unexamined assumption that both Newton and the wave advocates held in common, that the frequency, in the case of Newton the number of times in a given period that like points in a flow pass a point in reality, or the wavelength, in the case of the wave advocates the distance between like points in a wave, was ordered by the way that light comes out of a prism.

To see that white light isn't made up of all colors, we have to see that light is actually a frequency that is reduced when it interacts with matter to produce colors, and that when it interacts with the matter that makes up a prism, the resulting colors are not the order of the frequency. We have to re-order our thought to view pigments not as something that absorbs all colors except the color we see but rather as something that reduces the frequency of the light that strikes it so that we see a different color.

Frequency is the number of times a point on a flow passes a plane. Light is emitted in packets from the oscillating surface of combusting matter. The light travels at a uniform speed regardless of frequency so that the number of packets that cross a plane in a given time determines the light's frequency. Hot matter that burns with a high oscillation produces more packets than matter that is burning at a lesser rate. Thus, high frequency light is light with a high number of packets per time period when compared to lower frequency light.

Intensity is the amount of light contained in the packets at any point. Because light diminishes with distance, intensity diminishes independent of frequency. Intensity and frequency are equal at the surface of combustion, with the hotter the object producing the light, the higher the rate of oscillation, the higher the frequency and the more light there is in each packet.

The eye evolved to detect intensity. Intensity allows an organism to locate itself within its environment because, as light diminishes uniformly from its source, all points in the environment off of which light has reflected can have their relative distances determined by the resulting intensities of the resulting light.

Frequency, on the other hand, only contains information about the quality of the surface off of which the light has bounced. And one of the primary consequences of frequency, color, is not information at all.

The eye evolved two sets of detectors to navigate in the two environments it encountered, nighttime when all intensities are low and daytime when all intensities are high. Intensities are visibility, luminosity, and generally deal with the amount of light available. The detector that evolved to see low intensity light at night would also see a range of low frequencies. The detector that evolved to see high intensity light in the day would also see a range of high frequencies. The eye contains rods that detect low intensity light and as a by-product, low frequency light, while the cones, bulbous to cushion the higher intensity light in the daytime detects high frequency light as a by-product.

Color is a by-product of frequency. Color is the sensation of frequency on the detectors that evolved to detect intensity. The eye evolved to locate the organism to which it is attached in the environment by comparing the individual intensities of the light reflecting off of objects in that environment. As light bounces off of those objects, its frequency is altered. The rods detect low frequency light in a range of colors, blue to green. The cones detect high frequency light in a range of colors, red to yellow. This is why there are four primary colors, and why one of those colors, yellow, can be produced by adding lower light frequencies, green and red.

Frequency does not provide information about location, and survival techniques that evolved to take advantage of the eye's ability to discern frequency, mating plumages, camouflage, food identification, did not contribute to the evolution of the eye. While an organism can evolve characteristics that take advantage of frequency, organisms have no way to take advantage of the eye's ability to add frequencies. In nature, light reflects off of matter providing a single frequency. Light may refract, produce the colors of the rainbow, but organisms don't navigate by the light of rainbows.

Man, however, can produce pigments that duplicate the colors of nature by trial and error determinations of how frequencies from specific light sources are reduced by bouncing off the pigments. Once the highest frequency pigments are identified, they can be mixed to produce pigments of lesser luminosity. Positioning bright pigments can also produce many hues. The three color printing system of positioning dots of yellow, cyan and magenta can result in the eye seeing the surface as if the paint had been mixed, as a single frequency.

As in Newton's proof that white light contains all colors, flows of light can be added in the eye. Even though we produce the technology that permits the eye to add light, and thus the eye would not have evolved to see added light, it is this process of adding light that has resulted in the many misconceptions made by empirical science, misconceptions about white light containing all colors, pigments therefore absorbing all colors but the reflected color, and the eye having individual receptors for each of the primary colors, red, green, and blue, yellow considered a mixture of red and green to reduce the number of receptors, and about the reverse order of light's frequency or wavelength, with blue high frequency, high energy, short wavelength light and red low frequency, low energy, long wavelength light.

In order to demonstrate that white light does not contain all frequencies, that color is the result of white light changing frequency as it interacts with matter and that white light is the highest frequency light, followed by yellow, red, green and blue, we will examine how light's luminosity follows frequency, how the eye functions when it color adapts to frequency, and how adding frequencies produces not only the colors we see from adding light, but the colors we use to mix pigments.

The world doesn't operate on the basis of how a triangular piece of glass conducts light. The glass used to fabricate a prism might well be fashioned by lightning from a thunderstorm striking a sandy beach, and the water droplets left after the storm might well refract light into the colors of the rainbow, but light didn't come into existence to be refracted by a prism and the eye didn't evolve to see the light from a rainbow. If we want to know about the environment in which we evolved, then we have to confine ourselves to what exists in the environment, rather than monkey seeing rainbows and then monkey saying that physical reality is like a rainbow, we can conform it to what we see coming out of a prism.

Of all the mistakes that we have made in the four and a half centuries since Copernicus put the Earth where it belonged, moving around the sun rather than at the center of the universe, none boggles the mind like the mixing up of the spectrum. Against all physical and observational evidence, classical physics assigns blue light the highest frequency of any light while quantum mechanics, ever the obsequious mendicant to assumed reality in its relentless pursuit of the counterintuitive envisions blue light as a product of the highest energy level.

Luminescence, the quality of light that allows us to see it, is all out of whack with the scientific fact of wavelength as ordered by how light emerges from a prism. Any casual observer would equate wavelength with luminosity because the more there is of something, the more apparent it should be. Blue, being the least luminescent would therefore have the fewest waves, the lowest wavelength. We don't have blue fire engines running around our streets because blue is the least visible color in the spectrum. We started painting our fire engines red because red was on the other end of the spectrum. But even red was not the most luminous color. Yellow is the most visible color in the spectrum and thus all of our fire engines are now painted yellow.

We know by simple sight alone that yellow should have the highest wavelength, followed by red, green and blue. Even when we abandon the fantasy of wavelength in favor of the more realistic frequency, the highest frequency should produce the brightest color, the lowest, the dimmest. If we didn't scientifically see all of the colors in white light, we might even see white light as the highest frequency of all. But because we are taught and thus see white light as a mixture of all colors, we don't even attempt to compute the wavelength or frequency of white light.

By simple sight alone, by using the organ to which the subjectivity of color is assigned, we should know that white, yellow, red, green and blue light are the order of frequencies with white light the highest frequency and blue light the lowest frequency. However, empirical science limits us to seeing light the way it comes out of the prism, either red, yellow, green and blue or blue, green, yellow and red, depending on the theory of refraction used, and currently concludes that because blue light refracts more, and because refraction is inversely proportional to speed, blue light has the highest frequency and red the lowest.

If the fact that we don't paint our fire engines blue is not a sufficient clue that yellow and red light have higher frequencies than blue light, then consider color adaption in relation to yellow and blue light. When we look at a particular color long enough, our eyes become adapted to it. The best demonstration of this fact is the anomaloscope used to test for the failure to see red or green. We all know that red and green light combine to produce yellow light. This is one reason we see light as waves, the logic being that because yellow light comes out of the prism between green and red that we see the average of the green and red light waves as yellow light waves.

An anomaloscope has a white standard disk used to reflect colored light. Yellow light is reflected off half of the disk, and green and red light are reflected off the other half. A person can look into the anomaloscope and adjust the amount of red and green light until the resulting yellow light matches the reflected yellow light. An assessment of the extent to which the eye fails to see red or green can be made by measuring the relative amount of green or red light it takes to match the yellow light.

However, an eye adapted to either green or red that looks at both halves of the yellow disk at the same time uniformly sees the color to which the eye didn't adapt, green if red adapted and red if green adapted. It sees the non-adapted color uniformly even though only one side of the disk is created by combining green and red. This forces the conclusion that even though the reflected yellow light is not made up of green and red light, it acts the same as if it were so therefore it must be.

If the process of recombining colors on a white standard screen wasn't conclusive proof that white light contained all colors, the anomaloscope outcome was.

This is no proof at all, however, because a person that fails to see either green or red light can still see yellow. If a person can't see red or green, and yellow is made up of red and green, how can he see yellow?

This is another clue that yellow light has a higher frequency than red light.

The current concept that waves of red and green light combine to produce an average color of yellow pretty much accounts for how empirical science explains color adding of light frequencies. If, however, instead of averaging waves, we look at adding different colors of light as adding frequencies which accumulate in the receptors of the eye that evolved to detect intensity, then we can make sense out of color adaption, luminescence and color adding.

We are dealing with light that comes in five colors, white, which is not really a color but rather a range beyond color perception, yellow, the highest frequency the eye can see and still see color, followed by red, green and then blue, the lowest frequency that the eye can detect before it ceases to detect color. We want to see how adapting to one color produces another color when we look at still a third color. We also want to see how adding these colors together produces all the colors of the spectrum, as well as white light. And we want to do both consistent with luminosity and the pigments derived for color mixing.

To accomplish this, we can assign arbitrary numbers to colors that can be used to represent the color's frequency. Because frequency correlates with luminosity, the higher the frequency, the brighter the color, we should be able to manipulate these numbers so that the order in which we see luminosity is in accord with the reality of color adaption and color adding, the accumulation of colors in the eye as a result of the eye receiving simultaneous flows of different frequency light.

While the numbers that we assign to the colors are arbitrary, once assigned, they must consistently produce what we see in reality. Thus, by assigning the number "one" to blue light because of its low luminosity and the number "nine" to yellow light because of its high luminosity, and "ten" to white light because it is beyond the range of the eye to see color, we would expect to be able to color adapt to blue, look at white and see yellow both in reality or by the numbers, white, ten, minus blue, one, equals yellow, nine at the same time that we would expect to add blue light, one, to yellow light, nine and see white light, ten.

We will see in Chapter Ten how yellow light travels the least distance in a prism to produce the highest frequency. Red and green light fall out on either side of the yellow light in a prism which means that they are approximately the same frequency, with red moving the shorter distance through the prism and therefore having the higher frequency of the two. Because green and red are somewhere in between blue and yellow, we can assign them the numbers of "four" and "five" respectively.

With the numbers assigned to white light and the four color frequencies and the understanding that light moves in packets, ten packets of white light pass a point during the same period that one packet of blue light passes the point. During the same period, four packets of green, five packets of red and nine packets of yellow light pass the same point. The numbers, while arbitrary, do represent the relative frequencies of light with ten, white light, being the highest frequency and one, blue light, being the lowest frequency.

The eye can accumulate individual frequencies and can still see individual colors so long as the frequencies do not exceed the highest color frequency the eye can see, yellow, or nine. When the frequencies entering the eye total nine, then the eye cannot discern individual frequencies.

When an eye is color adapted, it can only discern color frequencies that can be added to the frequency the eye is adapted to. The higher the frequency the eye is color adapted to, the lower the frequency becomes that the eye can discern.

Thus, a color adapted eye has blinded itself to the frequency of the color it is adapted to. An eye that is color adapted to red is blocked from seeing frequencies below red unless it looks at a higher frequency, and then it only sees the frequency between what it has adapted to and the higher frequency. When the eye adapted to red looks at yellow, it can only see the frequency of yellow after the frequency of red has been subtracted, or the frequency of green. Yellow is nine, red is five and green is four, so that a red adapted eye, five, looking at yellow, nine, sees four, green. In this case, where the frequency is high enough to effect both receptors, the high range receptor is forced to see green because even though the frequency viewed is nine, the red receptor is blinded to frequencies up to five.

The reverse applies if the eye is color adapted to green with yellow, nine, minus green, four, producing red, five, but for a different reason. The rods are more widely dispersed than the cones, but in all but the periphery, the two are bunched together. The eye evolved to see areas of shifting intensities, areas controlled by Fechner's law which requires that for the eye to distinguish two intensities, their ratio has to be at least five percent, with one at least five percent brighter than contiguous areas. A corollary to Fechner's law is that the frequency one receptor is experiencing will suppress a similar frequency in the other receptor. When the eye is color adapted to green, which makes the entire eye a single area, both the rods and the cones are adapted to the green frequency even though the cone can not "see" the green frequency. Thus, when the eye is green adapted and looks at yellow, it sees yellow minus green, or red.

The reason that green and red light make yellow light is simple addition, green, four, plus red, five equals yellow, nine. When the eye is seeing yellow light, it is seeing the highest color frequency before the level of frequency blots out the eye's ability to see individual frequencies. As a result, the eye is not seeing individual frequencies of red and green light, but rather the highest frequency color before it ceases to see color at all. Yellow light is unique. It can be seen both by adding colors, green and red, and as the highest single color frequency. This explains why a color adapted eye sees both sides of the anomaloscope disk the same and why an eye that can't see green or red can still see yellow.

If blue light were a higher frequency than red, as is currently believed, then when the eye became color adapted to red it would be able to see green or yellow when it viewed blue because green and yellow are intermediate colors if viewed in accordance with the way light emerges from a prism. In fact, an eye adapted to red cannot see green or yellow when viewing blue.

However, when the eye is yellow adapted, it does see blue when it views white. Why? Because white light is the highest frequency of visible light, ten, and subtracting the frequency of yellow light, nine, from ten leaves us with the frequency of blue light, one. This can be reversed by color adapting the eye to blue light. Such a color adapted eye looking at white light will see yellow light, which is white light, ten, minus blue light, one, equaling yellow light, nine.

The relationship of blue to white is complicated by the fact that each is a bookend of color vision. Increasing the rate at which something burns causes frequency to increase from blue to red to yellow to white, and sometimes white light becomes so intense that we see blue. The blue color results from the fact that both white and blue are on the outside range of color perception and the eye is tricked, confusing the high range with the low range, both leading to an absence of color perception.

And, if we color adapt to white, what do we see?

Nothing. We are snow-blind.

Luminosity and color adaption demonstrate that the frequency of colors is not ordered in the monkey see, monkey say way they emerge from a prism, but rather from blue light with the lowest frequency, through green, red and yellow to white light. The so-called complimentary colors, blue and yellow, green and red, are merely complimentary with respect to the frequency they add up to and are embodied in the relationship of the ranges of the two receptors in the eye, with the top of the low range, green, and the bottom of the high range, red, adding up to the top of the high range, yellow, and with the addition of the bottom of the low range, blue, to the top of the high range, yellow, adding up to a range beyond color perception, white light.

Color adding, the accumulation of individual frequencies in the eye, also demonstrates that blue is the lowest frequency of color perception.

If we have a series of projectors capable of casting a color on a white standard screen, we can add and subtract colors in order to see how the sum total of the frequencies entering the eye equals the color the eye sees bouncing off the pigments in reality. Casting a red light on the screen will cause the frequency associated with red light to bounce off the screen and enter the eye. The eye registers the frequency by the number of times the frequency passes the plane of the eye's receptors. If we next cast a green light on the screen, we have two discrete events occurring on the screen. The frequency of the red light is bouncing off the screen and returning to the eye and the frequency of the green light is bouncing off of the screen and returning to the eye. When the green light bouncing off the screen enters the eye, it increases the frequencies entering the eye by the number of times the new frequency is passing the receptors in the eye, as has been seen with green light, four plus red light, five equaling yellow light, nine, the maximum number of frequencies that can pass the plane of the eye and have the eye still discern color.

With red and green light frequencies passing the plane of the eye so that the eye sees yellow light, what happens when blue light is added? The number of frequencies that are passing the plane of the eye is already the maximum number the eye can see frequency and still see color. With the addition of blue, the number of frequencies now moves beyond the eye's ability to discern color and the eye can only see white light. By adding blue light, one to yellow light, nine, the number of frequencies passing the plane of the eye has increased to the point that the eye can see no color at all, but rather can only sense the presence of light, white light, ten.

When the eye sees yellow light produced by green and red light, it does not see either the red or green as colors. However, when colors are added so that the resulting frequency is below the frequency of yellow light, the eye can simultaneously discern the individual frequencies, and the simultaneous process of discerning the individual frequencies begins to enable the eye to add frequencies so that we see the hues that mimic the myriad frequencies we see when light reflects off objects in physical reality.

While the various hues that light is colored in progress from low frequency blue to high frequency yellow, they do so over two paths, the paths controlled by the fairly equal frequencies of green and red.

One path starts with blue and increases frequencies by adding green until one bright point in the spectrum, cyan, is reached at the point that the intensity of blue and green light is equal. Subtracting blue reduces the frequency in this path until only green light is left. Adding red to green starts to increase the frequency of the green path until yellow, the highest color frequency, is reached.

The other path starts with blue and increases frequencies by adding red until a second bright point in the spectrum, magenta, is reached at the point that the intensity of blue and red light is equal. Subtracting blue reduces the frequency in this path until only red light is left. Adding green to red starts to increase the frequency of the red path until yellow, the highest color frequency, is again reached.

The two frequencies that are added must together be below the frequency the eye records as yellow or the eye will not be able to record the frequency as an individual color frequency. As long as the combined frequencies being added is less than the frequency the eye records as yellow, the eye can record both frequencies as their respective colors. This is why there are two distinct series of hues that can be produced by adding green and red to blue. When green is added to blue, the eye still discerns the green color. When the red is added to blue, the eye still discerns the red color.

Yellow, of course, is the third bright point in the spectrum, cyan, magenta and yellow being the points of departure for color mixing in the printer's trade, where light has to reflect off of pigments and thus pigments used in color mixing have to produce the highest frequency before they can generate lower frequency hues.

In color adding light, yellow would be useless because it is already the highest frequency. With light, frequencies are added rather than reduced. Thus, red, green and blue light, as in the typical television screen, are necessary to mimic all the hues in nature.

In mixing pigments we produce a pigment that reduces frequency so we start with the most luminous colors. In color adding light, we have to add frequencies and thus start with the least luminous colors.

When we add frequencies, we are seeing two frequencies at the same time. We have to be careful that we are not talking about combining intensities. A color television screen is composed of a series of pixels, picture elements, containing blue, green and red dots. A stream of electrons is fired at these dots and the dots emit the light in blue, green or red frequencies. For instance, if the electron stream hitting the green dot is very small, then the green light that will be emitted is a green frequency of low intensity light. The intensity refers to the amount of light in each emitted packet of light. If the electron stream hitting the green dot is increased, then the intensity of the green light being emitted is increased. There is more light in each emitted packet of light, but the frequency of the packet remains unchanged. It is by varying the intensity of the electron stream that hits each pixel composed of blue, green and red dots that the various hues we see on the screen are produced.

When white light passes through red, green or blue filters, it has its frequency altered by passing through the filter. A higher frequency is being reduced to a lower frequency. The light we see is the frequency of the color we see. When an electron flow strikes a colored dot on a television screen, a flow with no frequency is being converted into a flow with frequency, the frequency of the color the dot emits. If the electron flow is increased, the colored dot emits higher intensity light at the same frequency.

Other than the fact that intensity diminishes with distance and frequency doesn't, there is no difference between five flows of blue light each with an intensity of "one" and one flow of blue light with an intensity of "five". There is a difference between one flow of blue light and five flows of blue light, and a flow of blue light with an intensity of one and a flow of blue light with an intensity of five. The difference is in the area of the eye impacted. The eye, with its myriad of rods and cones, has to have a way to deal with the hard edges of reality off of which light is bouncing. It has to be able to see those hard edges. Fechner's law describes how it does this, separating areas of vision by the relative brightness of the light received in those areas.

When color is added in the eye, then, it is added in areas where the result produces a differential in brightness from contiguous areas. To show how flows of light add in the eye, we will duplicate the intensity produced by an electron flow striking a colored dot in a television screen with five flows of blue, green and red light. This will result in the numbering system being increased by five, from one to ten, to five to fifty. While we are using the larger numbers, which assume a larger area of the eye in which color adding is going on, any number can be divided by five to get the hue's value on our basic scale of one to ten. For instance, five flows of red and four flows of green light have a value of forty-one, (5x5)+(4x4), which, divided by five gives a hue value on the scale of one to ten of 8.2, almost yellow. We divide by five instead of the number of the flows being added to preserve the color adding feature of the example. Dividing by the number of flows is a process that results in color mixing.

We don't have to go further in the example and add blue, green and red simultaneously to see how frequencies add in the eye because adding each with the other covers the entire range of hues. For instance, it is clear from Newton's proof that equal intensities of each color produce white light. Lowering the three intensities in equal amounts drops the color through gray to black, the nonvisible frequency below blue. Varying the three intensities simultaneously merely fills in the spectrum found by reflecting light off of pigments in reality, colors such as the pervasive browns and exotic purples.

Finally, the fact that the eye has two receptors, one for the blue/green range and the other for the red/yellow range does not affect color adding. We have to remember that we are the ones that are using technology to take advantage of what basically is a by-product, the ability to see frequency as color, of the characteristic, the eye, that evolved to locate the organism to which it is attached in the environment. The lower range receptors easily add the colors with low luminescence, blue and green. If the limit of that receptor on our scale of one to ten is four and a half, midway between green and red, then most of the blue/green range of hues is added in that receptor. The exception is cyan, whose hue value on the scale of one to ten is five, meaning that it is perceived by both receptors. Compare its relative brightness to blue, with that of magenta, where both receptors perceive the entire blue/red range. Cyan is brighter when compared to blue than magenta. The addition of the high range receptor in cyan makes it clearly brighter relative to the hues in the blue/green range when compared to magenta's brightness relative to the colors in the blue/green range that use both receptors over the entire range.

With blue light assigned a frequency of one, green light a frequency of four, red light a frequency of five, yellow light a frequency of nine, and white light, ten, the frequency at which the eye cannot discern color, five flows of blue light have a frequency value of five (5x1). When green light is added, the sum of the frequencies crossing the plane of the eye's detectors begins to increase. With green's frequency four, adding five flows of green, a flow at a time, to the blue frequency of five raises the frequency in increments of four as follows: 9, 13, 17, 21 and 25. When five green flows have been added to the five blue flows, there are ten flows made up of five blue frequencies and twenty green frequencies (5x4). If the blue flows are removed one at a time, the frequency is reduced in increments of one as follows: 24, 23, 22, 21, and 20, which is five flows of green light. We started with blue light, mixed green light with it so that the frequency increase causes the hues to change, rising until we reach an equal mixture of blue and green, which is cyan. By removing the blue light, we reduce the frequency so that the hues change until the light is totally green. Notice that cyan has a higher frequency than either green or blue. This accords with the luminosity of cyan, which is brighter by far than blue, and somewhat brighter than green. This is why cyan mixes pigments better than either blue or green.

Adding red light to blue light produces an entirely different set of hues because the eye can still discern the individual colors. Maintaining the same numbering system, with blue assigned the frequency of one and red assigned the frequency of five, five flows of blue light still have a frequency value of five (5x1). When red light is added, the sum of the frequencies crossing the plane of the eye's detectors begins to increase. With red's frequency five, adding five flows of red, a flow at a time, to the blue frequency of five raises the frequency in increments of five as follows: 10, 15, 20, 25, and 30. Note that in the example, twenty-five is the same frequency as five blue and five green flows of light and will be the frequency of five flows of totally red light. This emphasizes the point that as long as the frequencies being added total less than the frequency of yellow light, they retain their character as an individual frequency and thus retain their hue. When five red flows have been added to the five blue flows, there are ten flows made up of five blue frequencies and twenty-five red frequencies (5x5). If the blue flows are removed one at a time, the frequency is reduced in increments of one as follows: 29, 28, 27, 26 and 25 which is five flows of red light. We started with blue light, mixed red light with it so that the frequency increase causes the hues to change, rising until we reach an equal mixture of blue and red, which is magenta. By removing the blue light, we reduce the frequency so that the hues change until the light is totally red. Notice, as in the case of cyan, that magenta has a higher frequency than either red or blue. This accords with the luminosity of magenta, which is brighter than blue, and somewhat brighter than red. This is why magenta mixes pigments better than either blue or red.

Notice also that the frequency value explains why cyan and magenta are closer in luminosity to green and red, respectively, than blue.

With the hues of added light between green and blue, and red and blue covered, we are left with the hues between green and red. Five flows of green start with a frequency value of twenty (5x4) while five flows of red start with a frequency value of twenty-five (5x5). The frequency of yellow is forty-five, five flows of nine (5x9) so that there can not be a color of green and red that is forty-five that can retain its individual hue. Yellow light, of course, made up equally of green and red light, is the equivalent of cyan between blue and green and magenta between blue and red, the hue of red and green with the highest frequency and thus the most luminosity, the brightest.

The process of adding and subtracting green and red is the same as with the blue and green series, and the blue and red series to complete the basic hues obtainable by color adding all combinations of two of the colors together.

Maintaining the same numbering system, with green assigned the frequency of four and red assigned the frequency of five, five flows of green light have a frequency value of twenty (5x4). When red light is added, the sum of the frequencies crossing the plane of the eye's detectors begins to increase. With red's frequency five, adding five flows of red a flow at a time to the green frequency of twenty raises the frequency in increments of five as follows: 25, 30, 35, 40 and 45. Five red flows have been added to five green flows of equal intensity, which is yellow (5x9), the limit of the eye's ability to discern color. If the green flows are removed one at a time, the frequency is reduced in increments of four as follows: 41, 37, 33, 29 and 25. We started with green light, mixed red light with it so that the frequency increase causes the hues to change, rising until we reach an equal mixture of green and red, which is yellow. By removing the green light, we reduce the frequency so that the hues change until the light is totally red.

The only thing left to do is to add the blue light to the yellow light, one to nine, to get ten, white light, the frequency past which the eye can discern color. This is accomplished by adding one to 45 five time as follows: 46, 47, 48, 49, and 50. The yellow light becomes whiter with each addition of blue light until the resulting light is white light, light beyond the range of the eye to discern color.

Of course, there are a gadzillion different frequencies so that to cover the entire color spectrum, we would need to start with a sixth of a gadzillion flows of each color. This might make the example overly long, however, when five flows can accomplish the same thing.

The way we have available to analyze the colors light comes in, its luminosity or brightness, the way the organ that discerns color, the eye, adapts to it so as to discern other colors in afterimages, and the way adding light produces hues as well as the luminous color we use to mix pigments, shows not only that colors and their hues are discrete frequencies, but that the color frequencies are ordered with blue light being the lowest, yellow light the highest, and green and red light being somewhat intermediate between the two. All three areas of analysis support the view that white light, contrary to Newton's unexamined assumption made three centuries ago that white light was not a frequency, but rather a bundle of all frequencies, is itself a discrete frequency above the eye's capability of discerning, and that all color hues are a result of this frequency being lowered as it interacts with matter.

Other than the fact that the ten or fifteen billion dollars spent looking for a blue light emitting diode on the wrong end of the color spectrum might have instead fattened stockholder pockets and reduced consumer prices, why do we want to know that the color we see is the result of white or yellow light frequencies bouncing off of physical matter in a process that merely reduces its frequency?

How we visualize light, how matter produces it and how it interacts with matter, determines how we visualize that matter, what it is and how it is constructed. And it is awful difficult to understand light unless we understand the matter that produces it.

If the color we see is the result of reflected frequencies, then white light is not made up of all colors. But because we think we know that white light is made up of all colors, when we see light strike different objects in reality and produce different colors, we have to ask, where did all the colors go?

Because we believed white light contains all colors, when we saw white light produce single colors in reality, we were forced to conclude that what we saw in reality was what was left over after pigments had absorbed all other colors.

Thus, a simple question, why does matter reflect different colors, became an extremely complex question, how is it that when light containing all of the colors in the rainbow strikes different matter, only one color is reflected? What happens to all the colors that aren't reflected?

The question of color, then, became the question of where the light we don't see went, an apt parallel to seeing no light on a collection screen and seeing water waves.

White light was all colors, so that when we saw one

color, we were seeing what was left of light after the surface had absorbed all of the other colors. Pigments, then, became something that absorbed light, rather than something that did what they obviously did, reflect light at different frequencies.

Instead of pigments being matter that altered the frequency of light so that the frequency moved from higher frequency white or yellow light to lower frequency colored light, the pigment that reflected red light was viewed as absorbing the green and blue elements of the white light, the pigment that reflected green light was viewed as absorbing the red and blue elements of the white light and the pigment that reflected blue light was viewed as absorbing the green and red elements of the white light. Because red and green light produced yellow light, a situation in which light is added, it was felt that blue pigment also absorbed yellow light even though with pigments we are dealing with light striking matter in physical reality rather than adding colored light.

Further, filters did not alter the frequency of the light so that yellow filters produced yellow light, red filters produced red light, green filters produced green light and blue filters produced blue light. Filters were instead absorbing vast quantities of light. Thus a yellow filter absorbed blue light (the obverse of blue light absorbing green and red light, which is combined into yellow light), a red filter absorbed blue and green light, a green filter absorbed blue and red light and a blue filter absorbed red and green light which included the yellow light into which they combined.

When first we decided that it took force to keep an object from moving in a straight line, Newton's first law of motion, we reversed the reality we recall in our minds, the reality we use to see the reality in which we exist. We didn't start to get everything backwards, we started to get everything exactly backwards, one hundred and eighty degrees away from reality. This resulted because objects don't require force to keep from moving, they require force to keep moving.

Once we had accustomed ourselves to seeing the reverse of reality, we went on to produce all sorts of conclusions that were diametrically opposed to the reality in which we exist and which we can physically measure. We have swirling masses of gas conglomerating matter rather than doing what swirling masses do, spinning matter apart. We have stars which burn from the inside out rather than matter burning the way we know matter to actually consume itself in reality, from the outside in. We have an explosion, the big bang, bringing matter into existence rather than doing what we know explosions do, tear matter apart.

In the field of color, instead of pigments creating color, we have pigments eliminating color!

If white light contains all of the colors that exist, then the only way we could see different colors would be for the matter that is different colored to be removing all of the colors from the light except for the colors that we see.

The question of color became, then, not what is there, but what isn't there.

You can't get any more opposite than this!

Instead of light bouncing off of different matter in the environment, and in the process, having its frequency lowered so that an eye that might happen by picks up the specific frequency and senses it, light strikes matter, and the matter starts to systematically absorb the gadzillions of different frequencies that could exist in the universe and be discerned by the evolved eye, allowing just a single frequency out of the gadzillions or so in the light to bounce off of it and strike the eye.

And because there are gadzillions of hues in the environment, many right next to one another, then the matter must be extremely busy, with one portion of the leaf absorbing all frequencies but the frequency that produced green, the portion next to it absorbing all of the frequencies except the one that produced red, the portion next to these two absorbing all of the frequencies except the portion that produced yellow, and so on, with orange and rust, ad infinitum.

Applying this dictate to the printer's profession, the creation of the bluish cyan wasn't a mixture of blue and green, it was the exclusion of red. The reddish magenta wasn't a mixture of blue and red, it was the exclusion of green. And yellow wasn't the mixture of red and green, it was the exclusion of blue.

Because white light breaks down in a prism into all the

different colors, and all of the different colors of light can be combined in the eye to produce white light, the only way that we could see individual colors in the environment is if all the colors but the individual color had been absorbed out of the white light by the object the light is striking.

Even though this is patently absurd, even though it is impossible to picture the gadzillions of different hues that light comes in all traveling in the same white light ray, even though it is improbable to think that some process going on in the sun or a light bulb is producing a million known frequencies and a gadzillion unknown frequencies, even though it is obvious that the prism and the matter in the environment are simply progressively lowering the frequency of white light because of the way that the white light interacts with the material that makes up the prism or the matter in the environment, nevertheless, pigments absorb light because white light produces colored light in a prism.

If we don't believe what the empirical monkey says, we can go monkey see colored light enter one side of the prism and come out the other side in colors.

It's absolute proof that white light contains all colors!

This is not going to be the first time we run into a basic conceptual difference between the consensual view of light and the way we will find light to be structured. As long as concepts give empirical science something with which it thinks it can use to measure the physical reality around us, then the concepts underlying those measurements are of little consequence. The fact that the end result of the concept has nothing to do with the concept, as here, where how a prism breaks down light has nothing to do with how light interacts with matter in reality, doesn't stop empirical science from claiming that the concept is proven by the measurement. We have for centuries been conditioned to say, look at the concept of mass/gravity, a straight line force which cannot produce circular motion in matter, it describes with absolute precision exactly where all of the matter in the solar system will be in a billion years, and where it all was a billion years ago, when in fact the only way we know where the planets will be tomorrow is the fact that we measured them there a year ago tomorrow.

Modern science is a confusing compilation of misconceptions that produce a Gordian knot of immense complexity, but the mental detritus that has accumulated on the conceptual tangle that is the current view of light is almost impossible to unravel. To do so, we are going to have to take it a thread at a time.

Knowing that white light does not contain all colors is one clear thread.

Knowing that the interaction of light with matter reduces its frequency is another clear thread.

Knowing that frequencies are ordered from blue as the lowest to yellow as the highest is still another clear thread.

Knowing that white light is a frequency is a fourth

clear thread.

The next thread we have to clear from the tangle is the absurd proposition that light is a wave.

3. The Horrid Analogy

The one thing that Newton got right, the idea that light is made up of particles, Newton's corpuscles, empirical science buried so deep that even Einstein's Nobel winning discovery of the photoelectric effect couldn't completely disinter the corpse, leaving the concept of light today as some sort of wave/particle produced by an energy shift of orbital electrons in an atom, electrons that exist as a cloud of possible probabilities whose myriad potential positions only become realities upon observation, a truly incomprehensible picture that carries with it its own justification, the empirical claim that for modern science to be comprehensible, for it to obtain the truth about the universe, the truth itself must be incomprehensible.

Which says more about empirical science than it does about Newton, whose biggest error, the mass/gravity concept, was not his error at all, but rather the result of an attempt to prove an inverse square law with respect to planetary rotation that failed. Even on his deathbed Newton was begging Halley, the Astronomer Royal at Greenwich Observatory remembered primarily for the comet named after him, to hurry up and get accurate measurements of the moon's orbit to prove his claim that his corpuscles, the amount of matter in the Earth and moon by volume, predicted the inverse square law. It was only a generation later, after it was discovered that Newton's corpuscles predicted nothing, that his disciples decided that the inverse square law didn't need proving, it could just be used to determine the amount of matter in the middle of planets.

The fault lies not in proposing explanations but in blindly accepting them as fact, a fault that can only be practiced by followers. With Newton's corpuscular theory of light, we have the curious case of the disciples destroying the proposed explanation and then when it re-emerged as a measurable fact that disagreed with the assumption, perverting the explanation to the point of incomprehensibility.

If light were made out of particles, Newton reasoned, then many of light's characteristics could be explained. Light as particles could explain why light traveled in straight lines, why it cast shadows, why it reflected off of mirrors and why it bent when it entered glass. Light as particles, however, did present one difficulty. If light was made up of particles, then how could it retain its coherency when one beam of light passed through another beam of light without the intersecting particles bumping into each other and destroying the light in both beams.

This is an exceedingly interesting question. We see objects in reality because light from the sun, light which is expanding over the surface of an expanding sphere, bounces off of points in that reality, the hard edges of reality. At every point that the light bounces off of the hard edges of reality, it produces a new expanding sphere of light. As we look around, we are able to discern tens of thousands of hard edges in reality because the expanding spheres of light, which are diminishing uniformly with the distance over which they expand, carry information about the relative locations of these points.

Somehow the expanding spheres of light from each point retain their coherence.

Looked at from a different point of view, we can line up ten radio stations in a row and start them broadcasting at different frequencies. We can then take a tuner, a device that can detect individual frequencies, and tune in each frequency in turn or tune in all ten with ten tuners. We can do this in a room with a light bulb shining overhead and the sun pouring in the window.

How is it that frequencies retain their coherence?

Because light, to a great extent, retains its coherency, Newton's colleagues thought that his corpuscular hypothesis wasn't too bright. Light couldn't be made up of particles because if we crossed one beam with another, we should eliminate both beams. Empirical science has a way of proving these things. Cross two beams, they both retain their coherency, ergo, light is not made up of particles. Shine light on a finger puppet, the corpuscularians responded, and the finger puppet casts a shadow, ergo, light is made up of particles. Drop two rocks in a pond and the resulting waves will interact, and in doing so, will retain their individual coherency. Light was therefore a wave phenomena.

An obvious solution to the impasses would have been to sit down and come up with some more concepts because it was obvious that neither concept explained all of the facts, and concepts are only as good as the facts they explain. An alternative would have been to say, light might be a particle, here's the evidence, and it might be a wave, here's the evidence, but golly, we just don't know, so we'll admit we are going to have to deal with concepts, and keep our eyes open for new concepts that will explain more better.

Empirical science, however, is the process of measuring things in physical reality, and then classifying them. We can have fish and we can have fowl, but if it is neither fish nor fowl, then its nothing. Because empirical science defines the categories in which it does the classification, a general consensus can be made about where to fit something. Categories can be expanded or created and characteristics by which things are classified can be downplayed or ignored. Thus, when empirical science has something lying on the table in front of it, still flapping or flippering, it's going to get in one category or another, its going to be classified as fish or fowl or its going to be filed away and forgotten.

Nothing is more unforgettable than light, unless it's falling objects, so light has to be classified. The problem is that in trying to understand the structure of light, we are not trying to classify something, determine its characteristics so that we can group things that have similar characteristics with it. We are trying to find out what it is and how it works, what generates it, what moves it, how it's held together, what happens when it gets to where its going. The process involved in attempting to comprehend the nature and structure of light has nothing to do with the process that classifies a mayfly as a fly rather than as a water bug with extraordinarily exotic mating habits.

Light is not something that can be classified. It is something that has to be understood.

Empiricists, eschewing comprehension in favor of mindless measurement, don't see this, and those engaging in the corpuscular/wave controversy were not concerned with understanding what light was. They were only concerned with how to classify it, because for an empiricist, classification *is* understanding. Thus, when Young came along with an experiment whose outcome predicted that light was a wave time and time again, on a more regular basis than mass/gravity predicts the circular motion of all of the matter in the universe, light was classified as a wave and its nature declared to be understood.

In fact, knowledge of the nature and structure of light could not have been known at the time Young performed his two-slit experiment that proved conclusively that light was a wave because electricity had yet to be produced in a sustained enough flow to allow the measurement of inductance. The people that were arguing about what light was were totally ignorant about the most important phenomena in physical reality. These empiricists, pretending to uncover objective laws that controlled the heavens above and the Earth below, were actually only creating concepts that masqueraded as universal truths and they were doing so in virtual ignorance. They didn't have enough information about reality to propose concepts that had any chance of matching reality. It is the concepts of these same men, the ignorance that is the legacy of dead men who knew nothing, objects travel in straight lines, gravity is a property of mass, planets rotate because of angular momentum, the solar system condensed out of a swirling mass of gas, white light contains all colors, light travels like water waves, light is not made up of what gives it off, that dictate how we see physical reality and cripple our resulting comprehension.

Young's proof that light was a wave involved the empirical practice of looking at nothing and seeing something. To see something where nothing exists requires a lot of argumentative discussion which allows the mind to become accustomed to seeing what is not there. The anti-corpuscularians had bolstered their water wave argument, that water waves can cross and still retain their identity, with the observation that sound waves produce the same effect. We can listen to an orchestra and identify the individual instruments, listen to the chorus and still hear the fat lady sing.

Waves retain their coherence. Therefore, light must travel in waves.

Sound waves, of course, need some sort of medium to travel through. They are produced by creating a vibration in matter. If we take a hammer and hit an anvil, we'll hear one big clunk as the physical matter of the hammer collides with the anvil. If it is a very good anvil, the effect of the hammer striking it might resonate though the physical matter itself, causing the anvil to actually sing. If the atoms of the anvil were air, they would be moving rapidly in front of the force created by striking the anvil.

However, with the anvil's atoms held in place, they simply move back and forth as the force passes them by. Cymbals demonstrate this effect on the surface of the matter because the surface has been made large with respect to the force. Bang two cymbals together and they resonate externally. The result is that the surface of the cymbal itself starts the air around it to vibrating, and we here the sound very clearly, perhaps too clearly.

The vocal chords vibrate like the cymbal, but the force is air passing through them from the lungs, and sound is the result of the chords flapping together.

Sound, then, is physically affecting the air around its source, causing the air to vibrate as the sound progresses through it. The vocal chords can flap all they want to, but if there is no air around to conduct the sound, then there is no sound.

The ear, on the other hand, is designed to detect the vibrations in the air and turn those vibrations into electrical impulses which are given meaning by the mind's recall system. If there is no ear in the forest, there are still detectable vibrations if a tree falls.

We have to see this very clearly if we are to understand the nature of light and the compounded mistakes empirical science has made in its attempt to come to terms with its inability to form a workable picture of light. Sound waves are produced by vibrations which set up oscillations in the air, physical movements of the atoms and molecules of atoms that make up the air. These movements extend in an expanding sphere around the source of the sound because there are atoms and molecules of atoms in every direction from which the vibrations emanate. The movements can be blocked, as with a brick wall, and they can be focused, as with a megaphone. But their transmission depends on the physical contact of the atoms and molecules of atoms through which they are having an effect.

The atoms and molecules of atoms that make up the air aren't themselves moving anywhere. They remain in position, vibrating back and forth as the wave passes them by.

Waves don't exist independent of the medium because they are a disturbance of the forces holding that medium together.

The production of waves in water is no different. There are always forces at work holding matter together. In the atmosphere, forces in the nuclei hold atoms and molecules of atoms together, while the force of the Earth's attraction keeps the atmosphere pressed down against the Earth's surface. When we are at the bottom of that atmosphere, flapping our vocal chords, we are producing a force that primarily disturbs the force holding the atoms and molecules of atoms that make up the atmosphere together. With water waves, we are contending with the force of attraction that is holding the water to the surface of the Earth. If we scream for help ten feet underwater, the vibrations don't travel very far because the forces holding the nuclei of hydrogen and oxygen atoms together are so strong that the vibrations of the vocal chords do not affect their relative positions. The force is quickly absorbed by the forces holding the water molecules together.

While earthquakes might be strong enough to send sound-type disturbances through solid water, resulting in a tsunami, an oversized tidal wave when it reaches shallow water at a shoreline, waves in water are basically a surface phenomena. First, a container is necessary. An ocean basin, lake bed or swimming pool is sufficient. The water must be at rest with respect to the forces acting upon it, in this case the force of the Earth's attraction, gravity. Next, of course, there must be water to be disturbed. Then follows a force which can disturb the water. A force sufficient to overcome the force of gravity has to be exerted on the surface of the water in order to disturb the surface of the water.

Finally, the surface of the water has to be bounded by atoms and molecules of atoms held together by a force less than the force holding the molecules of water together. If we cover our swimming pool with Plexiglas, we are not going to be able to produce waves on its surface because the forces holding the atoms and molecules of atoms of the Plexiglas together are stronger than the forces holding the molecules of water together.

When water's boundary condition is the atmosphere, then forces exerted on the surface of the water can overcome the forces holding the atoms and molecules of atoms of the atmosphere together, and the result of the force, waves, can rise into the atmosphere. Water waves, then, are the result of a force disturbing the force of the gravity maintaining water in a stationary position and the continuous production of waves is the result of the gravity gradually regaining the upper hand over the force that causes the disturbance, much as a bouncing ball slowly bounces to a halt.

The force of disturbance creating the water waves forces the surface down at the point of impact and the water around the point of impact is pushed out, and, because the atmosphere has less force, up. The contiguous water eventually overcomes the force of the impact and pushes back, causing the surface to rise at the point of impact. The surface falls back, producing another impact and sending another wave after the first. This process continues until the impact site once again comes to rest with respect to gravity. The waves move through the water until the force of gravity, and the forces holding the molecules of water together, can no longer be affected by the force of the original impact. However, the molecules of water aren't moving anywhere. They remain in position, bobbing up and down as the wave passes them by.

Visualizing sound and water waves has allowed us to form a clear picture of waves as a disturbance of matter by force which causes the atoms and molecules of atoms that make up the matter to move with respect to existing forces so that the effects of the force, the disturbance, can be detected at a distance.

Nothing is moving from one place to another!

Sound and water waves do not exist if there is no medium because waves are merely a disturbance of a medium and have no independent existence of their own.

The question of whether light is a wave or a particle, then, is not one of viewing light as a bunch of dots or a squiggly line, it is a question whether light exists at all!

Just as the assumption that white light contains all colors dictated the conclusion that matter absorbs all colors in order to reflect the color we see, the conclusion that light is analogous to a water wave or a sound wave dictates that light is neither emitted nor absorbed, but rather is a disturbance in a medium.

If light isn't emitted or absorbed, then it is merely a local phenomena, something that happens in the atom. If we use our fist to cause a wave, nothing in our fist has moved into the environment. If we are floating in the river when a motorboat roars by, our subsequent bobbing has no effect on our bodies. We don't obtain anything from the waves as they pass us by, except perhaps a little nausea.

Categorizing light as a wave or a particle two centuries ago and decades before electricity as a sustainable flow was discovered and inductance was observed would nonetheless determine whether in fact the light that illuminates our reality, the light that allows us to know that reality exists, itself existed. If light was a wave, then it existed only as a disturbance of a medium, a medium which was not measurable but which had to be there nonetheless because the fact that light was a wave dictated its existence.

If light was a particle, on the other hand, then light could be emitted from matter, exist as it moved through space and have an effect on what it struck.

If light was a particle then light could, in fact, exist.

With everybody's attention focused on whether to categorize light as a particle or with the much more measurable water and sound waves, it didn't take much to tip the scales in favor of categorizing it with readily measurable phenomena. These was little doubt that a science based on mindless measurement would choose measurement over meaning.

The empirical scientific method places supreme importance on the order in which facts are discovered rather than on any attempt to reformulate concepts that will incorporate new facts. And while the empirical scientific method demands that only unknown facts be used to prove hypotheses, when facts demonstrate hypotheses, it is neither necessary for them to have been predicted, nor for them to have been unknown. Young's two-slit experiment certainly gave the appearance of demonstrating the fact that light travels like water waves, at least to those who had become accustomed to seeing things that didn't exist in reality, scientists driven by the need to please patrons and to all the rest that have followed, quiescent souls bowled over by reputation, authority, and the general pressures of the consensus, opting for tenure, contracts, fame, book deals and television appearances over comprehension.

Young's two-slit experiment is a simple process that divides light and allows the divided light to intermingle and strike a collection screen. The result of the experiment is that the light that strikes the collection screen appears as bars of light and no light. If the light isn't divided, a normal amount of light leaves the source, expands over an expanding sphere, diminishes as it does so, and then falls on the screen according to the inverse square law of diminishing light. When the experiment is performed, when the light is divided and intermingled, the light that falls on the collection screen is the same amount of light that would have fallen on the screen had the light not been divided. The only difference is that, instead of the light being uniformly distributed over the surface of the collection screen, it is distributed in bars of light and no light, with the bars of light being twice as bright as the light would have been had it been distributed uniformly over the collection screen.

Somehow, Young's two-slit experiment results in the divided light being recombined so that the light that strikes the collection screen strikes it in concentrations that are twice as bright as the light would have been had it not been divided. The extra brightness comes at the expense of the dark areas, and these dark areas are areas out of which light had been recombined with the other light to form the bright bars the experiment produces on the collection screen.

Even monkeys seeing the results of this experiment would have had to monkey say that light was being recombined, and in the process, the light falling on the collection screen was falling in bars of recombined light, and bars of no light out of which the recombined light had been recombined. A close examination of the physical set up of the experiment would have verified this conclusion.

The physical set up is extremely simple. All that is required is a card that doesn't conduct light with two identical slits cut in it side by side so that when the card is placed between the light and the collection screen, the light is divided by the two slits and allowed to intermingle after it passes through the two slits.

Following the light in this physical set up, we can see it being emitted from its source in the form of an expanding sphere moving out in all directions in space. The reason that light diminishes inversely with the square of the distance over which it travels is the geometric relationship of the myriad expanding spheres light forms as it moves away from its source of emission. A sphere has an area of four time the constant π times the square of the radius. An expanding sphere of light has a common source of emissions, the surface of whatever is doing the emitting. For every sphere, in fact, four and π are constants, with the only variable being distance, the distance to the source of emissions, or the radius of the sphere, the surface of the expanding sphere of light. Because the area of a sphere is four π times the radius squared, the area of any sphere defined by expansion from a common source is related to the square of the distance to the point of emissions. As there are any number of points on a radius, especially when we are dealing with spheres with surface thicknesses of one light particle, all of these spheres are related to one another by the square of their distances from the source of emissions.

As an expanding sphere of light moves away from its source of emissions, the surface area of the sphere gets larger in proportion to the square of the distance from its source. As the surface area gets larger, light evenly distributes itself over the increasing area. Light expands into the area available for its expansion. Because the area available for its expansion is increasing with the square of the distance to the source of emissions, the light expanding over that surface is diminishing at the same rate, the square of its distance from the source.

This is why light contains information. While there would be no way for a detector such as the eye to evolve that could tell how far it was from the source of the light, because it would have no way of knowing how strong the light was that was being emitted at the source, such a detector could evolve that could take information from various sources of light, specifically the light that was bouncing off the hard edges of reality, and use the relative intensities of the flows to locate those hard edges of reality in the environment and thus the detector's position with respect to those hard edges. This is how the eye evolved to locate the organism to which it is attached with respect to the reality in which it evolves.

By placing the double-slitted card between the light source and the collection screen, the experiment alters the center of expansion of the expanding sphere of light, creating two new expanding spheres of light with new centers of expansion out of identical sections of the original expanding sphere of light. As the light passes through each slit, portions of the original expanding sphere are given a greater area in which to expand. With new centers of expansion, new expanding spheres begin to expand from each slit. Smaller parts of the original expanding sphere are being allowed to expand over entire hemispheres. As a result, the light, expanding into the area available for its expansion, expands faster than it would have had it remained as a part of the original expanding sphere of light.

After the light passes through the two slits, then, two identical expanding spheres of light are expanding over hemispheres whose centers are lined up on identical points a short distance apart. The resulting expanding spheres of light intersect at all points of their expansion. This produces flows of light from each sphere that are parallel with flows of light from the other spheres throughout the intersecting areas of expansion, flows of light that are congruent. The recombination of these parallel flows occurs throughout the expanding hemispheres at each point of their expansion, and the resulting recombined light falls on the collection screen geometrically as bands of bright light and no light.

The bands of no light on the collection screen are areas out of which the light has been recombined into the bands of bright light.

If one slit is used, and the slit is twice the size of the two slits used to produce the experiment, the amount of light hitting the screen will hit it uniformly and will equal the light contained in the bright bands when the double-slitted card is used to provide the expanding light with new centers of expansion.`

If the experiment is designed so that the distance of each slit from the light source can be altered, then moving one slit a distance that is different from the other slit changes the center of expansion, and in the process, the area over which the light is expanding. This in turn changes the rate at which the light from one sphere is diminishing, altering the light's intensity. With the individual intensities of the two expanding hemispheres of light no longer identical, the parallel flows do not recombine, leaving the light to fall on the screen in the same manner as if it had passed through a single slit equal in size to the two separate slits.

What is recombining the light?

With the empirical scientific methodology requiring predictive facts to prove concepts, absolute importance is placed on the order in which facts are discovered. Known facts cannot be used to prove concepts (unless the consensus agrees that they can) and once a concept has become scientific fact, facts in reality that don't agree with the concept aren't facts. As a result, empirical science turns concepts into scientific facts which exclude the application of later discovered facts.

A corollary of this methodology, then, is certainly that the order in which we discover facts dictates the extent of our ignorance. Joseph Henry demonstrated inductance in the eighteen twenties, but by that time light was already traveling in incontrovertible waves as a result of Young's two-slit experiment.

Induction is the force that occurs at right angles to a flow of electricity in a conductor. If you point your right thumb in the direction of an electrical flow, then your fingers will be curling in the direction of the inductive flow. If you take two electrical conductors containing flows moving in the same direction, then the inductive flows will combine the conductors in much the same way that the lines of force around a magnet can affect the physical positioning of the magnets.

Young's two-slit experiment was performed approximately thirty years before Henry demonstrated induction, and as a result, Young was totally ignorant of the possibility that something might be recombining the flows of light in his experiment. Had induction been considered by Young, he might well have concluded that the results of his experiment, the bars of light and no light, was caused by inductive flows associated with the passage of light, the same inductive flows that could recombine physical conductors, recombining the flows of light to the extent that those flows were compatible.

He could then have answered the question why flows of light from separate sources intersect without destroying each other. He could have stated with some force that the reason that light was coherent was that it was held together by inductive flows and that light was therefore most certainly the result of the movement of particles of matter.

Not knowing anything about inductance, however, he concluded that the bars of light and no light on the screen were identical to the effect that occurs when water waves from two sources intersect with each other.

If we drop two identical pebbles onto a completely calm pond surface several inches apart, the water waves will move out in concentric circles, interfering with one another. At points where the crests of each set of waves intersect, the waves reinforce each other, as they do at each point at which the troughs intersect. However, where a trough intersects with a crest, the two cancel each other out exactly and the surface of the combined waves at that point is identical to the surface of the undisturbed water.

Young analogized these points at which the two waves canceled each other out to the dark bands on the collection screen, and the bands of light and no light were called interference patterns, mindlessly encasing what was actually going on in ignorance. As the experiment proved conclusively that light traveled in waves, moving the slit to another distance had the effect of moving the waves out of phase so that the waves, notice it is no longer light, but waves, the waves then fell uniformly on the collecting screen. By measuring the distance between the slits, and the distance of the slits from the screen, the wavelength could be determined, always a necessary outcome of the empirical process which is really only looking for something else to measure regardless of whether that something exists or not.

The failure of mental acuity in accepting Young's two-slit experiment as proof that light travels like water waves is in the analogy itself. Forget that water waves need a force to be disturbed. Forget that the force has to have something, water, in which it can express itself. Forget even that light is the three dimensional propagation of expanding spheres and water waves take place on a two dimensional surface.

Forget all of the obvious reality.

When intersecting water waves cancel each other out, there is still something there, water.

When the light hits the screen, there is no light.

Nothing is there!

This takes a little dwelling on. Empirical science is only good at showing us things that don't exist in reality, or, if they do, are incapable of independent measurement. In a process that claims to be able to prove hypotheses, concepts, to be facts, the only hypotheses that survive are those that cannot be proven not to be fact, which stand the test of time in the parlance of the consensual delusionists. Thus, Newton's mass/gravity, failing to predict the inverse square law by computing force based on the demonstrable volume of matter, turned the computations around to use the inverse square law to determine the amount of matter in a given volume, how much matter was at the center of the Earth or the Sun, results which could never be independently verified. Using motion to create matter, empirical science has produced the modern marvels of dark matter and black holes, fantasies which, because we evolved to see matter by the way it emitted light or reflected light, are forever beyond our senses and thus our ability to measure.

Empirical science continually asks us to see things that aren't there because it is incapable of explaining the causes for the things that are right in front of our eyes, what makes objects fall, what makes the planets rotate on their axes, what makes them orbit the sun. This method of misdirection, widely practiced by magicians with all sorts of silly stuff up their sleeves, finds its highest expression in getting us to look at a collection screen upon which there is not a scintilla of light, not a shred of anything, and saying that what we don't see is the same thing we see when water waves intersect, that in fact when we see nothing it is just like seeing something, water.

It is a horrid analogy, made even more horrid because dumb us, wide-eyed children that we are, allow it to pass unquestioned.

Young's two-slit experiment tells us one thing and one thing only about light, that light is made up of flows of particles and that the same inductances that are measured with the passage of electricity and which form the basis of modern civilization are associated with the flows of the particles that make up light.

This one thing in turn tells us a lot about the physical reality for which we are attempting to construct a consistent picture. It tells us that the light that fills our environment is actually structured, a cohesive flow of particles that are held together by their forward motion and the inductances that their presence produces. It tells us that light is actually something, not just some mysterious disturbance of some medium in which we exist, that we can understand it, and that if it is something, that it comes from something.

The question whether light is a wave or a particle belies the magnitude of the ignorance answering the question that light is a wave produces. It's not a question of whether light goes around obstacles or whether it casts shadows, it's a question of whether light actually exists. Waves have no existence independent of the medium through which they are propagated, and to claim that a wave can act like a particle when we measure it acting like a particle, but is otherwise a wave is to express an ignorance so profound that only the empirical process of monkey see, monkey say could perform it without embarrassment.

The simple fact is that if light is a wave, then it is a local effect like gravity, with no existence other than its effect, and thus beyond analysis.

The horrid analogy creates a picture of light as a local disturbance with no independent existence outside of matter, a picture that totally obscures the more realistic picture of light as something produced by something happening in matter so that matter gives up some of what makes it up, and what it gives up physically moves from the matter and physically occupies space as it moves and affects other matter that it encounters as it moves.

Empirical science had the option of creating a picture of something that existed in reality or something that wasn't there, a ghost, and it did as it was accustomed to do. Unable to form a picture of reality that it could explain, it chose to present a picture of ghosts, pictures which can never be explained rationally, and therefore, never challenged.

Empirical science, by denying the existence of light, precluded light from being made up of what gives it off.

4. Light *Is* Made Up Of What Gives It Off

Our understanding of magnetism, electricity, heat and light is the result of a perception that gradually becomes aware of effects, assigns words that create a picture that explains each effect, and then uses the pictures the words paint to dictate all subsequent awareness that emerges with increased observation. As we became aware of magnetism, electricity, heat and light, we built up different pictures of each because each had different effects. Faced with different pictures of their different effects, but beguiled by their similarity and interrelatedness, we periodically attempted to unify the pictures, create a single picture of all of them. It is no secret that no one has been able to unify concepts dealing with theses four forces, the four most interrelated phenomena in our existence, phenomena which surround us all of our days all of the time every place we are, awake or asleep.

What is a secret is that no one is even trying.

With more important questions to ponder, questions such as how big the meteorite was that caused the extinction of the dinosaurs, how far the nearest black hole is and does it have a wormhole connected to it so that we can whiz through space/time to probe it, what is the texture of empty space and how does time distort it, it is generally assumed that we know everything there is to know about the phenomena that surround us, that we need only a little consolidation to come up with the formula that explains all of our existence.

When we know everything there is to know, there is no reason to question what we know about the obvious physical phenomena of our existence. Gravity as proportional to and a property of mass may not tell us what is going on when an object drops, and it certainly has no way of describing the rotation and orbiting of the planets, but we aren't rocket scientists, after all. Someone, somewhere must know the details, if not on Earth, then in heaven above.

We are so befuddled by our success in manipulating magnetism, electricity, heat and light to produce technology that we mistake knowing what happens when we twist a knob or poke a key with actually knowing what is going on. We ask questions, what is magnetic force, where does electricity come from, how does light come out of a light bulb, what is heat, but a science that seeks only to describe what is happening and shuns asking what is going on can merely respond with palliatives, magnetism is caused by molecular magnets, electricity is a moving charge, light is produced by a change in energy state, heat is motion.

As long as we tolerate a science which can hide its ignorance behind mindless, meaningless explanations, we will have a mindless, meaningless science.

Before we can penetrate the relationship of light to matter, we have a major conceptual block to overcome, as major as the ignorance produced by assuming white light contains all of the colors of the spectrum, or picturing light as ripples in a pond. We are blocked from seeing light as being made up of the same thing that the matter that gives it off is made up of, a truly enormous feat of empirical science, a feat akin to the astronomical practice of monkey seeing fiery comets and monkey saying snowballs.

Although the modern explanation for why light is not made up of what gives it off rests on Newton's conceptualization of mass to explain why objects fall, the process through which light was precluded from being made up of what gave it off actually started thousands of years ago under the classical concept in which the elements were divided into Earth, Water, Air and Fire. Fire, which is made up of heat and light, was considered to be the most noble of elements because it rose in Air. Air was considered the second most noble element because it rose in Water. Water and Earth both sought out the center of the Earth.

Earth and Water had weight and Air and Fire had negative weight.

In addition to the weight characteristics of the purported elements, Fire was something entirely different from Earth. When Earth, which was anything that wasn't Water or Air, was burned, Fire was liberated, rose into the heavens, and joined the fiery stars in whatever sphere they revolved.

The separateness of matter and the fire the matter produced, then, was ingrained in the culture by the time that we began to wonder just what was going on when something burned, when we began to realize that earth was more than one element and that air contained something other than just air.

When something is ingrained in our culture, it is invisible. Earth is earth, and fire is fire and the idea that what makes up the earth might be given off as fire is simply beyond our ability to formulate the question. There was an atomistic concept of matter, a hard ball of matter held together by the gravity that its mass produced. However, there was no concept of an atom with a nucleus and orbiting electrons that could be given off, and without an atom that could permit something to be given off, there was no way we could pose a question dealing with whether fire could be made up of what gave it off.

When we started asking what fire was, we were confronted with the question of whether fire and heat were the same thing. Before we could answer that basic question, we were immediately confronted with the question of the motion of heat, which clearly came from the fire and moved across space to affect the senses, and the fire itself which had motion, its movement in space and the observable fact that the heat departed the motion of the fire and moved across space. And then there was the fire's luminescence, the fact that the fire produced flames, and it produced heat that moved beyond the flames and it also produced light that moved beyond the flames.

And this wasn't two thousand years ago, or even a thousand years ago. It was only three hundred years ago, during the period that Newton was proposing that light was a corpuscle and others were arguing that light was a wave. This was only a hundred years before Young's two-slit experiment proved conclusively that light was a wave.

How empirical science has since dealt with heat and light, the two elements of fire, is the measure of empirical science. Light is not measurable, and certainly not quantifiable. Heat is measurable, but not as a component of light. As a result, empirical science, unable to measure light, turned it into something that doesn't exist. Able to measure heat on a limited bases only, empirical science proceeded to limit its knowledge about heat to what it could measure.

In both cases, empirical science ignored the obvious realities involved in heat and light. The first reality is that light proceeds from its source in an expanding sphere that diminishes inversely with the square of the distance from the source and in doing so, it diminishes out of existence after a point. The second reality is that light and heat are the same thing.

Empirical science could ignore these realities by removing light and heat from the definition of things, as something, as being a part of the matter that produced them, and as a result maintained the millennials old misconception of Earth and Fire as two separate things.

The difference between the classical concept of Earth and Fire, and the current concept of matter and light is that fire actually existed and light doesn't, while the heat component of fire exists only as energy, an amorphous concept which encompasses the measurement of work done or doable.

To see light as made up of what gives it off, we have to see it as something that moves from one place to another, and to see it as something that moves from one place to another, that exists between its source and its destination, we have to realize that we are not talking about two things, heat and light, but one thing, frequency. Because we feel frequency with two different detectors, one in the skin, the other in the eye, as heat and light, we are led to the erroneous but eminently empirical conclusion that they are different things.

Three hundred years ago the concept of phlogiston was created to explain the effect of fire. While the nature of phlogiston was as solid as our concept of light today, which is to say, totally incomprehensible, all things to all people and anything to anyone in accordance with what his desires are at the time, one thing was certain about phlogiston: It was something that moved from one place to another. When a fire burned, something came out of the fire, moved from the fire, traveled across space, and hit whatever was in its way, there to be taken up by that something and, for better or worse, to become a part of that something. Fire could heat up an iron skillet and thus the products of fire, heat and light, were something that moved out of the fire and across space to the skillet.

Of course, the characteristics of phlogiston were legion. Because phlogiston was fire, and fire was a noble element that rose in air, many people thought that phlogiston had a negative weight like the element it came from, fire. Thus, when it left burning matter, it made the matter heavier, and when it was absorbed by something like an iron skillet, the iron skillet became lighter. Others thought that because it was something that was moving from place to place, that the opposite should be the case. If it were in fact something, then it must have weight so that when it was lost in burning, it made the matter lighter, and when it was absorbed by matter, it made the matter heavier.

Lavoisier put an end to the argument about phlogiston by claiming to demonstrate that combustion was the process of atoms of fuel and oxygen combining, and, after the process was finished, the weight of the resulting product exactly equaled the weight of the atoms of fuel and oxygen that went into the process. Because the weight of the burnt remains equaled the weight of the remains before they were burned, heat and light were imponderables, stuff of no substance, simply the intangible effects of the process of chemical combination.

After Lavoisier disproved the existence of phlogiston as something that existed and moved from one place to another, in effect proved that light didn't exist, Young was able to demonstrate by his two-slit experiment that light in fact didn't exist, that it was merely a disturbance of a medium. The lack of a medium to propagate light was taken care of by drafting luminescent aether to the task, something as unobvious as the light that didn't exist was obvious.

Heat, however, was clearly tangible. It had to be disappeared another way. Instead of categorizing it with light, which we only see and don't feel, heat was classified with energy, with things that did work as energy.

Fire could produce heat which could boil water which could produce steam which could operate engines that could lift iron ingots. Horses could lift skids which were weighted down by iron ingots. If one and a half horses could lift ten iron ingots over one foot in one minute, and a steam engine could do the same, then the steam engine was rated at one horse power. If it took one ton of coal to produce the heat that allowed the steam engine to lift the iron ingots, then the amount of energy trapped in the coal could be computed as a horsepower equivalent.

Empirical science could measure all energy on a relative basis with respect to the work the energy could accomplish so that it could deal with energy quite well without asking bothersome questions about what it was or how it was produced. If someone got too curious about heat, what it was and where it came from, the curiosity could always be slaked by meaningless empirical laws, such truisms as energy can neither be created or destroyed, or, because a small amount of energy is lost in every transaction, the universe is running down. Empirical laws stand sentry to the search for knowledge, saying stop, this is the law, ask no further, protecting empiricists mindlessly from ever having to question what they know and thus exposing what they don't know. Once heat was categorized with things that did work, and things that do work as energy, heat simply dissipated.

It is fairly obvious that the laws dealing with energy can never be empirically proven. That's why they are laws, secrets of the universe purportedly uncovered by empirical observation. We see some physical effect in reality, we make up some sort of explanation for it, and if no one can successfully show that the explanation is contradicted in reality, the explanation itself becomes reality. No one has been able to demonstrate a connection between matter and gravity, nor find its cause in matter, but that is not necessary. Newton viewed planets orbiting the sun and came up with the idea that they did so because they would be moving in a straight line but for a force that was pulling them toward the sun, the force of gravity. Gravity as a property of and proportional to mass was a hidden secret of nature that Newton had uncovered to the eyes of all, the phrase actually being, once it is demonstrated, it is apparent. Unless the law can be empirically shown to be wrong then it will stand the test of time and be eternal. And because everyone now knows why objects fall, no one ever has to question why objects fall again, they fall because gravity is a property of mass.

Of course mass/gravity doesn't explain planetary orbiting any more than corned beef and cabbage explains why we get hungry. In a civil society matters have to be proven beyond a reasonable doubt. If I am accused of theft, I don't have to prove that I didn't steal. My accusers have to adduce evidence to demonstrate that I did. The need to prove a negative is only necessary in unenlightened societies where conviction is a requirement, because it is impossible to prove a negative.

But when it comes to heat and light, empirical science doesn't hesitate to prove negatives because heat and light are obvious realties. By separating heat from light, each can be declared nonexistent separately. By quantifying heat by what work it does, and then subsuming it as energy, it can be polished off by the great negative that energy can neither be created nor destroyed.

If something can't be created, then what is it? Nothing. But it can't be destroyed either, so it must be something.

This disposition of heat is fairly obvious. Eliminating it with mindless gibberish, we still have something to make us believe we know what heat is. It is the energy that pushes jet planes and boils our coffee water.

Light, on the other hand, presents a problem. Because light doesn't do anything but illuminate the universe, we can't convert it to work and mindlessly measure it away. We can convert it into colors, but that doesn't produce any work either. Color is just more nothing that can be used to generate more mindless laws, laws about white light containing all colors, blue light refracts out at a shorter frequency than red light, red light is a longer frequency than yellow light, or light doesn't exist between its cause and effect other than as a disturbance of a medium required to prove light's nonexistence.

As a result, proving a negative about light, that light is not something, that it doesn't move from one place to another, is not as simple as disappearing energy from the mental landscape. But prove it we must because while we might have difficulty viewing heat as a wave phenomena, light is clearly visible so that its very lack of effect, its nonexistence, is obvious. If we hold our palm up and grab what comes from the fire, our palm will hold the heat, but it won't retain the light.

Light's very apparentness makes its existence easy to deny because even though we see it all around us, it has no substance. But how do we prove nothing is there?

If nothing is there, then all empirical science has to do to confirm that light is, in fact, immaterial, an imponderable, that it is nonexistent, is to show that it does not have the same characteristic of weight that matter has.

If light has no weight, it is not matter, and if it is not matter, then it is not made up of what gives it off!

Why was weight determinative in deciding that light had no existence? When Lavoisier was creating the rule that combustion was the chemical combination of oxygen and fuel atoms, the only concept of the atom that was available was the concept of a hard little ball of indivisible matter, basically Newton's corpuscles, which had survived the shift to density, mass determined by motion, as the cause of gravity. As he put the oxygen and fuel atoms in an airtight glass container and allowed them to burn until the oxygen gave out, his thoughts were focused on the possibility of these little balls of gravity producing matter escaping through the glass. He had no concept of an atom that might contain something else, something that might be given off as part of the process of combustion. The concept of an electron was over a century away.

What is an electron?

In the eighteen seventies, Thomas Edison produced an invention that ran an electric current through a filament in a vacuum tube, producing light. Some twenty years latter, Thomson, working late into the night under one of Edison's electric lights, started to experiment with a tube invented by Braum. Using a vacuum tube which, instead of running a current through a filament, placed a gap in the electric circuit so that something came off one end of the gap and entered the other end of the gap, he started to examine what we now know as a flow of electrons. By placing a metallic sleeve in the vacuum tube, he could direct the flow of electrons through the sleeve, causing it to move out of the vacuum tube and affect objects at a distance. This predecessor to the television tube raised some interesting questions about the nature of matter.

At first, many observers thought that what was moving across the open circuit was the same thing that made up light. After all, why should changing a load to a gap change what moved through the circuit?

However, the flow could be deflected with a magnet and therefore had a charge. Light wasn't deflected by a magnet. Whatever was moving across the gap was therefore not the same thing as light. Light didn't have a charge.

Thomson decided that whatever it was that was moving across the gap was a negative charge, it was a particle, it was smaller than the particles that were thought of as atoms, and it was produced by different types of atoms so that it must be a particle that was associated with atoms regardless of the element the atom made up. Visualizing atoms as hard balls of matter, Newton's corpuscles, Thomson determined that whatever it was that was affecting objects at a distance, it was a negative corpuscle, and it was named an electron. It moved from one opening in the gap to the other because of the opposite charge of the other opening in the gap. Being a negatively charged particle, the electron had to be moving toward the positively charged opening in the gap.

It wasn't until the dawn of the twentieth century, then, that a concept of matter being able to give off what made it up even became possible. Using charge to distinguish what was moving across a gap in a circuit in a vacuum tube from what was moving away from a filament that was part of a circuit in a vacuum tube was premature, the result of thinking in terms of another made-up law, polarity. It's extremely difficult to visualize that the simple separation of a filament in a vacuum tube changes the nature of what is flowing through the filament, but these are the type of calls empirical science makes when what it sees is dictated by what it makes up.

Whatever an electron is, it was discovered to be a part of the atoms that made up matter and those atoms took on a complexity that would indeed permit matter to give off some of itself without appreciably affecting its weight, the empirical test Lavoisier chose to demonstrate light to be an imponderable.

Because empirical science finds it necessary to measure its concepts, an electron is thought of as having a mass that would allow almost a billion times a billion times a billion electrons to make up a gram. This is not quite a gadzillion, but it is getting there. Had Lavoisier known, when he was uncovering the immutable laws dealing with whether light was an imponderable or really existed, that something as small as an electron was associated with the little balls of matter that he had in his mind, and that billions of atoms might make up even the smallest portion of fuel, then he might have had second thoughts about passing a law that said that light wasn't made up of what gives it off. Even if he could have weighed an electron, and it's questionable whether he had scales sufficiently sophisticated to weigh what he claimed to weigh, the failure of the fuel to lose weight after it was burned, he wouldn't have been able to weigh the electrons given off as light, if in fact electrons are the same thing that makes up light.

But, pushing the envelope just a little further, just what was his concept of weight. When we say that it took a billion times a billion times a billion electrons to make up a gram, we are saying that an electron has a mass of one billionth of a billionth of a billionth.

What does mass have to do with weight?

Newton set out to prove his theory that gravity was a

property of and proportional to his little corpuscles of matter by taking the volume of the Earth and the moon, computing the amount of gravity there would be in both if gravity were proportional to the amount of matter, and then predicting that the amount of gravity produced by the two would equal the force it would take to divert the amount of matter that makes up the moon from moving in a straight line.

Needless to say, it didn't work out for him, the moon having too much matter to stay in orbit.

After Newton's death, things became much worse when the same logic was applied to the planets and found to be sorely lacking, with no object in space capable of being used to predict the inverse square law.

Not being able to use the amount of matter to predict the inverse square law, and therefore prove that gravity was a property of and proportional to matter, Newton's disciples decided that it wasn't the inverse square law that needed proving, after all the inverse square law could be measured simply by dropping a rock. It was the inverse square law that could be used to determine how much matter existed in the moon and the planets. This was a perfect empirical conclusion because it could never be checked. No one could ever weigh a planet or a moon or determine what was inside of them.

What empirical science was actually doing was creating matter by measuring motion. This is the source of black holes and the dark matter we can't see making up galaxies.

And the matter empirical science was creating was mass, the amount of matter that had to be at a location so that the gravity it created would produce the motion that was measurable, straight-line motion, the motion everything is assumed to have but for the existence of gravity.

The earthly analogy of mass is density, because with density, just as with the planets, volume is deceptive. Volume does not determine weight, density does. The more dense something is, the more it weighs for a given volume.

But, just like energy and the intensity of light, weight is a comparative measurement. Objects do not have intrinsic weight. An atom of hydrogen and an atom of iron will both fall in a field of attraction at the same rate. They will have no weight until something attempts to come between them and the source of the attraction. When something comes between atoms and the source of attraction, the force it takes to stop each atom from moving is translated into weight.

This is Galileo's famous experiment at the Leaning Tower in Pisa. The experiment shows that weight is not a product of the matter that makes up the atom, but rather is a result of the force it takes to either move an atom against a field of attraction, or the force it takes to keep an atom from moving in a field of attraction.

Weight is therefore a product of force on the nucleus of

an atom, and not a product of a force on anything that might be orbiting that nucleus. Thus, if the electrons that orbit the nuclei of atoms were released as a part of the combustion process, we wouldn't expect to be able to measure any weight gain or loss. The release of the electrons, with weight being a product of a force on the nucleus, would not have any affect on the weight of the nucleus.

The irony with using Newton's mass to conclude that light was an imponderable, that it was a disturbance of a medium, is that Newton's mass/gravity requires that objects move in Galileo's frictionless space, a notion Galileo made up in an attempt to prove that it was the Earth's rotation rather than the sun's motion that gave the sun the appearance of going around the Earth.

Using Newton's mass to prove that light was an imponderable, a wave phenomena resulted in the creation of a medium in space through which the light waves passed that would negate the frictionless space Newton needed to base his belief that gravity was proportional to and a property of mass!

This illustrates the total disregard that empirical science has for the reality that it seeks to measure once it has assigned concepts to what it has measured. Once the concept of white light containing all colors is assigned to what happens to light in a prism, we never see white light again, we never even try to determine its frequency, we simply see a bunch of colors even though what's in front of our eyes is white light. Once light is determined to be a wave , we never see light again, just processions of ripples in the universal pond. Once the concept of mass/gravity is assigned to falling objects, we never again see falling objects, we never even try to understand what is making them move, we see only what we can't see, density. We assign to matter the property of weight and fail to see that weight is merely a product of the resistance of matter to a force of attraction.

When it came time to make a determination whether light was made up of what gives it off, in effect, whether light was something, empirical science used its concept of mass to prove its concept of light as something that didn't exist even though the resulting light required a medium that would disprove the basis for the proof of mass.

How can empirical science today, with sophisticated methods of measurement at its disposal, still consider light to be something other than heat, and therefore perpetuate the idea that light is nothing, that it isn't made up of what gives it off?

How can light still be empirically stripped from heat?

Light emerges from a prism so that it appears that blue light refracts the most followed in succession by green, yellow and red light. Refraction can be explained as the process of wave fronts reacting to matter in accordance with the rate at which they enter the matter. The slower the wave is moving, the greater it will bend. Because it is a law of the universe that the amount of refraction is inversely proportional to its rate of motion, an eminently unmeasurable proposition, to refract the most, blue light must be slowing down the most when it enters the prism, proving it has the shortest wavelength.

We have seen that wavelength is actually frequency, and while in the classic concept of light as a wave, the shorter the wavelength the higher the frequency, in reality, the highest frequency that emerges from the prism is not blue light, but yellow light, with red light following.

When empirical science uses the term infrared it actually is talking about infrablue, violet being the range at which frequencies move below the eye's range of perception, and when empirical science uses the term ultraviolet, it actually is talking about ultrayellow, although there is, in reality, no range of frequencies above visible frequencies, with hot capable of only becoming hotter and whiter.

We can see this mistake, the misordering of frequency so that red light is the lowest and blue light the highest, much more clearly by using the empirical definition of heat, an accurate description of what heat is.

Empirical science defines heat as motion.

Thus, we would expect that the higher the frequency of light, the more heat there would be associated with that light. We would therefore expect little or no heat to be associated with light on the blue end of the spectrum and a lot of heat to be associated with light on the red and yellow end of the spectrum.

That, of course, is what we in fact measure. Heat is associated with red light and not with blue light.

However, because empirical science considers red to be the longest wavelength and blue the shortest, it sees heat associated with the light that is just becoming visible, red light, light which it believes to be at the low frequency end of the spectrum, and sees the heat gradually disappearing as the frequency of the light increases until the heat has entirely disappeared when light reaches what empirical science believes to be the highest frequency of the spectrum, blue.

As heat does not increase with empirical science's belief about how wavelength or the frequency of light increases, empirical science concludes that heat is something different from light!

Empirical science has created a law which dictates that light emerges from a prism by wavelength, wavelength being an inaccurate concept to start with, concludes that red is the least amount of light and blue the most and then measures the light's energy, or heat. When the light's energy, or heat, is measured and found to disagree with the way empirical science's laws dictate light's frequency or wavelength should be, and finds that energy, or heat, decreases as the light moves from red to blue across the spectrum, it concludes that what it is measuring, energy, or heat, has nothing to do with light.

In short, it has been empirically proven that heat and light are two different things.

And it can prove it any day of the week simply by looking at how light comes out of a prism and then measuring its heat content!

This bifurcation of frequency into heat and light has given rise to one of the most incredible displays of ignorance in modern science, one that has distorted our entire view of the universe. We have seen that light expands as it moves away from its source, and it does so in a measurable manner so that it diminishes inversely with the square of the distance from its source.

This means that light expands out of existence!

Although empirical science has no term for what happens to light that has expanded beyond its physical bounds, after all, there's nothing to measure when something actually ceases to be, and a science based solely on measurement has nothing to measure if something ceases to exist, (as opposed to measuring what doesn't exist to start with, the no light in Young's two-slit experiment), the word that describes what happens when light expands out of existence is dissipates.

Light expands over the surface of an expanding sphere until it dissipates.

A candle that can be seen twenty miles away, but which can't be seen twenty miles and one foot away, has expanded over the surface of an expanding sphere, diminished as it has done so, and dissipated when it has diminished to the point that it is no longer in existence.

A candle that is twice as bright as the first candle might expand over the surface of an expanding sphere for forty miles before it dissipates, but dissipate it will.

We live on a planet in a sea of space full of stars. We have no way of knowing how big the universe is, but the question contains its own contradiction. Space is nothing, and nothing is just that, nothing. When we ask how big the universe is, we are asking how extensive nothing is.

It is matter that occupies nothing, and asking how far matter extends in nothing only points out that it probably extends pretty far. The only way that we know there is matter in nothing is that the matter is undergoing combustion. When matter undergoes combustion, it produces light, and it is the light from the stars undergoing combustion that we see.

We don't see reflected light from planets orbiting other stars because reflected light expands out of existence, dissipating before we see it. In like manner, direct starlight expands out of existence.

Thus, there is a quantifiable distance from the Earth past which starlight would never reach us on Earth because before it got here it would have dissipated, expanded out of existence.

However, modern science actually believes that it can make stronger telescopes to move us closer to the source of that light so that we can see it before it expands out of existence. Basically, the telescope allows us to see the light that doesn't exist from the furthest reaches of the universe.

And, if empirical science speaks of the nonluminous portion of the electromagnetic spectrum it actually believes that it does not expand out of existence, but just goes right on forever!

Heat and light, the electromagnetic spectrum, are the same thing, they are frequency, with heat being the sensation the receptors on our body evolved to sense, and light being the sensations the receptors in our eyes evolved to sense.

In addition, light is something. Mass/gravity explains nothing, and certainly cannot be a bar to figuring out what light is. Light is not a wave/particle, it is something that is emitted during the process of combustion, it moves from one place to another, it does things along the way, and has an effect directly on the matter it bounces off of or penetrates.

In the process of emitting light, matter undergoing combustion is diminished because the stuff that makes up light makes up the atoms that make up the matter that is doing the combusting. Matter, in fact, is merely one manifestation of the elementary particle that makes up everything in the universe, with light, the electromagnetic spectrum being the other manifestation.

It is no accident that immersing matter in light causes the matter in turn to begin to burn and give off light.

The universe is really quite simple, with nothing, empty space, and matter, the elementary particles that make up light and matter. The matter in the universe operates in a cycle, with combustion the process by which matter is converted to light. The light expands over the surface of an expanding sphere until it breaks down into individual elementary particles.

If there are no emission fields about in the space where these expanding spheres of light dissipate, then the elementary particles that made them up reform into matter and the matter in turn participates in the combustion process.

Because the mechanism that moves matter in space, both rotational and attractional, is a by-product of the mechanism that regulates the uniform expansion of the light, all matter participates in the cycle, being attracted to combusting matter and becoming a part of that combustion process and its resulting expanding spheres of light.

Light is merely a range of frequencies visible to the eye, the total range of frequencies produced by oscillation, either by the combustion process, the process of going from hot to cold, or manmade or environmentally made oscillations in electric or magnetic processes.

There are definite characteristics about these frequencies: Each frequency is disparate, it does not travel as a part of any other frequency, a lot of frequencies are not contained in a single frequency; The frequencies are defined by particles making up a length that is determined by the rate with which the particles are released, by the speed of the oscillations occurring at their point of release;

and the particles that make up the frequencies are the same particles that make up the matter that is producing the frequencies, matter is giving them off in combustion or is directing them by electromagnetic means.

The question then becomes, what are the characteristics of these particles that both make up matter and the light by which we see the matter.

What can we use to synthesize a particle that could describe what we see in physical reality?

Before we can answer that question, before we can synthesize an elementary particle that can explain light and matter, we have to know something about the matter that produces the light.

And, as we found with light, we will have to get rid of a major misconception about that matter before we can understand how it is that matter produces light.

That misconception is polarity!

The assumption of polarity in the atom we use to make up the matter in our reality blinds us to the actual nature of that reality.

5. Polarity: Nothing Dictates Everything

While there are conceptual problems with understanding the nature of light derived from the concepts surrounding light itself, the concept that white light is made up of all of the colors that we see in physical reality, the concept that light travels as a wave with no substance other than its effect and the concept that light is not made up of the same thing that gives it off, one of the biggest conceptual hindrances to obtaining a clear understanding of the nature of light deals with the pile of misconceptions that have built up around what matter itself is. After all, if light is in fact made up of what gives it off, then we have to have a reasonably accurate picture of what gives light off before we can understand what light is. And if light is not made up of what gives it off, we still have to know what's going on in matter that produces the effect of light.

The source of light is matter, and our concept of what matter is made up of rests on a misconception so basic as to be even more invisible than the misconceptions dealing with light as all colors, light as waves and light as not made up of what give it off. It is more basic than the misconception that have objects moving in straight lines unless a force alters that motion and thus is tied up irretrievably with the misconception of what makes objects move when we drop them. In truth, it is more basic than the notion that stronger telescopes move us closer to a source of light so that we can see what otherwise would have diminished out of existence. It even surpasses our basic misconception that we can turn our misconceptions into fact by finding facts our misconceptions predict.

If we have a basic misconception about matter, then we truly know nothing!

The basic misconception we have about matter deals with the atom we have conceptualized to explain what matter is. This atom is constructed using an unnecessary concept, and this unnecessary concept has forever befuddled our view of physical reality.

The concept carried into the physical picture of the atom we have constructed is the concept of polarity, the creation of protons in the nucleus of the atom to balance the negative charge of the electron orbiting the nucleus. This concept slides from our original descriptions of magnets through our emerging awareness of inductance, heat and electricity to our present day concept of an atom made up of a bundle of unrealized potentials.

Taking a bar of magnetic material, what was known as lodestone, and breaking it in half allowed us to discover interesting characteristics about magnetic material. If we attempted to push the broken ends together with the unbroken ends, we found that the two ends would resist being connected. There was some sort of invisible force pushing the broken and unbroken ends apart.

On the other hand, if we attempted to push the broken ends together, or if we attempted to push the unbroken ends together, we didn't have to push. An invisible force would reach out and grab the magnets and cause them to come together by themselves. Once together, force was required to get them apart again.

Empirical science loves to describe things in the form of laws which become incontrovertible once they are verified by independent experimentation.

The empirical law dealing with magnets, then, would be that likes attract and opposites repel!

The likes in this case were the broken ends with the broken ends and the unbroken ends with the unbroken ends while the opposites were the broken ends with the unbroken ends.

Thus, when faced with the question, what is pushing the magnets apart and what is pulling them together, we can just say, it's the law, likes attract and opposites repel. The statement doesn't tell us anything about magnets, but it answers the question. Empirical science doesn't require that answers contain content, but merely that they be in accord with the measurements that gave rise to them.

After we have codified our law dealing with the observation of the behavior of magnets, someone comes along and notes that one end of a magnet, if allowed to swing free, always points in one direction, the place up where it's cold, north. If we paint a little "n" on the ends of the broken magnet that point north and paint an "s" for south on the other ends, then we find that our law no longer applies to magnets. If we break a magnet in half, then the ends that point north will be one of the jagged ends and one of the unbroken ends and the ends that point south will be another of the jagged ends and the remaining unbroken end. With the "n"s and "s"es now considered the like ends of the magnet because we are now defining like by what points north, if we try and push the "n"s together, the magnets resist coming together, while if we attempt to push an "n" together with an "s", then these two ends snap right together and require force to separate.

Thus, the empirical law with respect to magnets now becomes likes repel and opposites attract!

When we are faced with the question, what is pushing the magnets apart and what is pulling them together, we now say, it's the law, opposites attract and likes repel. The statement doesn't tell us anything about magnets, but it answers the question. Empirical science doesn't require that answers contain content, but merely that they be in accord with the measurements that gave rise to them.

When we ask, what happened to the old law, likes attract and opposites repel, we are told that we have witnessed a demonstration of how empirical science, by formulating hypotheses which can be independently verified by predicted facts, advances knowledge. Expanded observation expands understanding.

When we say that it seems that we still have the same mindless statement to describe the behavior of magnets, and have not increased our understanding of what is making the magnets behave the way they do, we will more than likely get the response that empirical science doesn't deal with philosophical questions. Asking why magnets attract or repel is like asking questions about God, and we are much too busy describing where the universe came from to spend time wallowing in philosophical claptrap.

It's difficult to believe that our concepts about what makes up matter are derived from our concept about what is "up" with reference to our geographical sphere of reference, but that is the case. When we decided to define magnets by which end pointed north instead of by which ends of a broken magnet attracted or repelled within their own sphere of influence, we called the end that pointed toward the North Pole of the planet the north pole of the magnet. With one end of a magnet pointed toward the North Pole, and therefore named the north pole of the magnet, it was only logical to name the end of the magnet that pointed in the other direction the south pole because the South Pole of the planet was on the opposite side of the planet from the North Pole. If this seems overly simplistic, we are dealing with empirical science which satisfies itself with merely naming things.

With the magnet now having a north pole and a south pole, and with the south poles and the north poles attracting each other and the north poles repelling the north poles and the south poles repelling the south poles, we had to come up with some sort of generic statement about all this attracting and repelling. What was it about the magnets that underlies the law that opposites attract and likes repel? What could this law be called? Well, because we had north and south poles, and both the North Pole and the South Pole had one thing in common, they were poles, we had a law dealing with polarity.

Once empirical science names something, it takes on a life of its own. All of a sudden, magnets had polarity, polarity was the reason that the ends attracted and repelled, and if we really wanted to get into the nitty-gritty of why the ends attracted and repelled, we could look deep into the magnetic material, discover mental molecular magnets, and simply say in a fit of esoteric realization that polarity was the result of the polarity of the molecular magnets that were lining up with their north ends pointing north and their south ends pointing south.

We still aren't saying anything intelligible, anything with information content, but we are obscuring the answer to the question of what makes magnets attract and repel to the point that the average questioner has to say, well, I have no idea what this guy is talking about, so he must be a lot smarter than I am and who am I to question someone that's so smart that I can't make sense out of anything he's saying. Because the guy talking polarity knows everything there is to know about how magnets operate, he surely isn't going to delve into the real reason magnets attract and repel, and because we are too stupid to even understand what the guy that knows everything about magnets is talking about, we certainly aren't going to try and figure out what is really going on when magnets attract and repel, and as a result we are all going to go along ignorant about one of the most obvious physical effects in our reality.

And not to flog a dead horse because the horse is anything but dead, but when empirical science is put on the defensive about the inanity of its statement that magnetism is the result of molecular magnets either being in alignment or out of alignment, it withdraws from the molecule to the atom, itself a concept, and when asked what it is about the atom that produces magnetism, it withdraws into the electron, an even more tenuous concept, having been created to replace the phlogistic-like effluvium used to explain magnetism and electrostatic charges clear down to the twentieth century, and when pressed about what it is about the electron that produces magnetism, empirical science shines forth in a sort of progressive regression that it is the electron's spin that produces magnetic effects and when braced to explain what it is about spin that accounts for magnetism, it replies that it is not the spin per se that produces magnetism but rather the uncompensated net spin, and this unbalanced spin can be quantified precisely as the magnetic moment of the electron.

But, we ask, aren't you just measuring what exists and then creating a bunch of mumbo-jumbo to quantify the measurement. You haven't got any more explanation for magnetism than you had when you were blathering on about molecular magnets. No, empirical science answers in a towering rage, the magnetic moment of the electron is exactly measurable and it equates precisely to the magnetic effects.

But, it explains nothing, we insist.

We're not philosophers, were scientists, the empiricists answer. We gather knowledge. Only God can provide meaning.

Polarity is so embedded in our concept of the atom that we have totally lost sight of how it got there. We are saying that the language concept we derived to locate the opposite geographical poles of the planet dictates how the atom that makes up matter operates. It's a ridiculous assumption but one that clouds every picture we try to form of the microcosm, be it how chemical interactions work, how our body works, or how the weather works.

For purposes of analyzing light, polarity is not a cloud, it is a block to forming any picture of what's going on because polarity limits the cause and effect of light, and thus its existence, to a mere local displacement of the electrons orbiting the atomic nucleus. To understand light we have to understand the matter that is producing the light and to understand the matter that is producing the light we have to have a rational picture of an atom that reflects reality.

The knowledge that started us down the road to assuming polarity in the atom is as old as history. The fact that rubbing an amber rod with a silk cloth would cause bits of paper to be attracted to the amber goes back to before the time that paper was paper and amber attracted bits of papyrus. With amber being rare, the effect didn't reach the status of a parlor pastime and thus incipient science until it could be performed with glass rods.

Gilbert, in his book De Magnete, published in sixteen hundred, tried to raise the parlor game to the status of scientific investigation. He undertook an examination of how effluvium, the magnetic moment of the time, reached out from the rod and grabbed the bits of paper. Today, the effect is known as electrostatic attraction and we have as much of an explanation for it as Gilbert had, which is to say, no explanation at all.

It's difficult to reconstruct the history of an idea concerning physical reality when the people describing the physical effect had no clear idea of what they were themselves describing. Gilbert, well aware of polarity in magnets, the fact that two ends repelled as well as attracted, could only find electrostatic attraction. Obviously, as soon as an awareness of Gilbert's analysis began to sink in, the fact that magnets attracted and repelled and rubbed amber only attracted, the question of the counterpart for repulsion in rubbed amber popped up. Where was electrostatic repulsion?

We had two categories and one was missing something. This set off a century long search for electrostatic repulsion.

The parlor games playable with the electrostatic effect were not limited to attracting bits of paper. If there was enough of an electrostatic effect, a spark and some crackle could be produced, much the same as we today shock our friends by scuffling our feet across a rug and then zinging them with the resulting charge. Magnets definitely didn't do this sort of thing, so it might have been wise to write electrostatic attraction off as something other than a magnetic effect. After all, magnets didn't attract paper.

However, we were dealing with invisible forces moving invisibly through space and causing matter to which the objects were in no way attached, to move. There was a lot of room for speculation. Objects fell without any apparent force moving them. The planets rotate and orbit without any apparent force moving them. A magnet could induce another magnet to move without any apparent force causing the movement, and we, with the hand of man, could take a glass rod which couldn't make anything move anywhere, and by the simple expedient of rubbing it, cause bits of paper to be attracted to it, to move toward the rod without any apparent force causing the movement.

In addition, we, with the hand of man, could cause the rod to produce a spark much like the spark we see during thunderstorms, the spark of lightning. One of the things that Franklin is credited with, his famous kite flying experiment, was designed to prove just that fact, that lightning was just like the spark that occurred when a glass rod was rubbed with silk and then brought in contact with another object.

One way that we could look at what was happening would be to say that rubbing the glass rod removed something from its surface. In doing so, it gave the surface a need to get that something back. If the surface was left at rest relative to its location in the air around it, the surface would gradually leach whatever it had lost back out of the air and return to normal. If the surface were brought close to something like bits of paper that had some of what it had lost, and that paper was light enough to move with whatever it had that the rod didn't, then the atoms of paper that had what the rod had lost would attempt to move toward the rod. And, if the rod had lost enough of whatever had been removed from its surface, and it was brought close enough to a surface with a lot of what it was lacking, what was lacking might just jump across the surface in the form of a spark.

Thinking along these lines, we might even say that lightning was simply the build-up of something in clouds that the Earth lacked for one reason or another, and when the buildup became greater than the clouds could sustain, whatever it was jumped through the gap of the atmosphere to the Earth.

In other words, the whole business could be explained with one charge, a charge which resulted in likes attracting likes so that when some of the particles were lost, the charge thus exposed would seek to attract them back. Electrostatic effects don't need polarity to explain them, they just need a charge that attracts like charges. There is no rational reason to bring magnets into the picture, just as there is no reason to bring gravity or planetary rotation, other examples of movement as a result of hidden forces, into the picture.

However, matter was made up of little balls, likes didn't attract, they repelled, and empirical science's need to categorize and analogize put finding electrostatic repulsion at the top of its agenda. In the first half of the eighteenth century it was proposed that there were actually two types of charges involved in rubbing stuff, a vitreous electrification that occurred when glass was rubbed and a resinous electrification that occurred when amber was rubbed. Both, with thinking immersed in fluidic concepts, were liquids, with vitreous fluid being positively charged and resinous fluid being negatively charged. Thus, an uncharged body held equal amounts of both vitreous and resinous fluid, and a body obtained its charge in accordance with which one was lost in the process of rubbing. A loss of resinous fluid would result in a positive charge while a loss of vitreous fluid would result in a negative charge, the loss of one leaving a surplus of the other. Complicated experiments can display a magnetic type repulsion between any similarly charged objects if the charge produces lines of force that can oppose each other. With the mental manufacture of positive and negative fluids that bought concepts of electrostatic charge into line with magnets, the stage was set for the injection of polarity into electricity.

If we ignore the spark and focus on attraction and repulsion, the magnetic analogies, then either gaining or losing vitreous or resinous fluids doesn't imply a direction of any sort. We are dealing only with the objects and the gain or loss of whichever type of fluid is merely a means to an end, a charge of some sort, charge being what was left over and what was causing the attraction.

Once Franklin's kite injected the spark into the discussion, the question of the fluid's movement became a necessary part of the picture of what was happening with electrostatic effects. The spark moved from one place to another, as did lightning, so the question became, what made the spark move.

Franklin concluded that, because opposites attract, the spark was made out of vitreous fluid and it moved to an area in which there was a lack of that fluid. An area with a lack of vitreous fluid was an area in which the charge was negative. Franklin therefore came to the conclusion that the whole shebang was merely the process of a positive charge being attracted to a negative charge.

Franklin successfully brought into play the concept of the electrostatic spark being *motivated to move* because it was being attracted to an area which had a charge opposite to the charge of the spark.

The mindless explanation for the movement of magnets, polarity, was applied to the movement of the spark!

We didn't have the foggiest idea what was causing the magnets to move, or even what magnets were. But then we didn't have a clue about what the spark was or where it came from.

It is well to pause on this picture, polarity as the motive force for a spark, before we see it carried further into a place where it has no applicability whatsoever, a circuit in which there is no to or from. One end of a magnet points to the north geographical pole and, even though when we use the pole to name the magnet's end we establish a law which says that likes repel, we assign that end the designation of north. The magnet has two physically opposites points, just like the planet, so its a convenient way to form a picture of attraction and repulsion.

Then we come along and find that we can create an attractive force by rubbing a rod. We say, well, the attractive force of the rubbed rod is like the attractive force between opposite ends of a magnet, so there must be a way to create a repulsive force by rubbing something. We think we succeed in doing that, proving that there is both an attractive and repulsive force associated with the process of manually making an attractive force and we then create fluids that mimic the effect of the ends of the magnet, with one a positive charge and the other a negative charge. Eliminate one and you are left with the other.

Focus on that.

We create two fluids to mimic the ends of a magnet and then produce charges by manipulating the imaginary fluid.

We produce a charge by eliminating the opposite charge!

When we then looked at the spark we could also create by hand, by rubbing the rod, we imbued the spark with the stuff rubbed off the rod. If the rod had two fluids of equal charge, then rubbing the rod either added one fluid to the rod, giving it more of a charge than the surrounding area, or it removed some fluid from the rod, leaving it with a charge opposite the charge of the removed fluid. Those that viewed the charge as receiving more fluid of one charge saw the spark jumping from the finger to the doorknob while those that viewed the charge as removing one of the fluids saw the spark jump from the doorknob to the finger. If anyone thinks this is a settled argument, visit the question of whether lightning strikes up or down in an era of slow motion cameras.

Franklin viewed the rubbing as removing the fluid, specifically the vitreous fluid, leaving the rod with a resinous or negative charge. He then said that the spark was composed of a positive fluid being attracted back to the negatively charged rod to once again bring the rod into balance with respect to its oppositely charged fluids.

For something to gain a charge, it has to lose an opposite charge!

Forget the fact that the search for electrostatic repulsion, vitreous and resinous fluids carrying positive and negative charges, and electric sparks actually being the movement of some sort of fluid were all just made-up poppycock.

The concept that opposites attracted was now embedded into any future concepts we might create about what makes up matter!

This is not small change.

Just as the notion that white light is composed of all colors dictated that the primary activity of matter was the absorption of colors, rather than the reflection of frequencies, the idea that the spark moved as a result of polarity dictated that the point toward which the spark moved had to have a charge opposite to the charge of the spark.

Here we had profoundly ignorant men, ignorant of even the most basic concepts about what electricity was, or what matter was, that knew nothing about batteries, generators, power lines, who were totally unaware of not only the present day concept of an atom, but any concept of an atom other than Newton's little corpuscles of matter which made up physical matter according to volume, who hadn't even learned that the light that made up the spark was actually nonexistent, a wave motion in a medium yet to be determined, we have these speculators into the beyond establishing concepts that would dictate how we would view light, electricity, matter and energy forever after simply because we mistake their concepts about the facts they discovered for the facts themselves.

And the concept that they were putting at the basis of all future analysis was the concept of polarity.

It is a concept that electricity, or light, needed a force to move it in the first place.

The physically measurable fact that movement is a property of the particle that makes up light, that after we slow down light, it speeds right back up again, and that the source of energy is the containment of that movement was excluded by the concept of polarity.

The force that moved electricity was polarity, the force that moved the opposite ends of magnets together, a force that was actually a charge when demonstrated by rubbing a glass or amber rod with a silk cloth.

If we don't have an opposite charge, then we can't get the spark! The opposite charge causes the spark. The spark wouldn't move without the opposite charge. The charge is what is moving the spark.

This closes completely the opportunity to view the movement of the spark as a property of the stuff that makes up the spark, released energy, with the charge being merely what gives the spark direction.

And to make matters worse, polarity also excludes the measurable fact that, if the planetary poles are removed as a frame of reference, that the spark could be viewed as moving toward a point having the same charge as the stuff that makes up the spark.

By mindlessly applying the rule we made up to explain how magnets acted when their behavior was dictated by classifying which ends pointed to the geographical north, that opposites attract and likes repel, to what was going on when we rubbed the rod with a silk cloth, we created opposite charges which had to get back together. We totally eliminated the obvious view of what was going on, that the rod had charges on a surface that was held together because of the charges, the charges were just that, a force inherent in the matter that attracted other matter with similar charges, in fact the surface was a surface because the charges were holding the matter that contained them together into a physical structure, and that when the cloth passed over the surface of the rod, it scrapped some of these charges off of the surface, leaving a deficit of charges on the surface. If the surface was brought close to bits of moveable paper, the deficit of a charge attracted the charges on the surface of the paper, and if the rod was brought close to something that couldn't move, but could give up some of its charge, the charge would jump the air gap and replaster the surface of the rod to make up the deficit.

In short, it is not opposites attracting, the physical concept derived from magnets, it is likes attracting, the simple electrostatic charges that anyone doing the laundry is more than familiar with, laundry whose surfaces after losing massive numbers of charges to the warm air of the dryer cling together to satisfy each other's lost charges.

Static cling is not the result of opposite charges, it is the result of like charges attracting.

We were, however, as a result of Franklin's perceptions of electricity as something positive that was attracted and moved as a result of polarity to an area where there was a negative charge, conceptually blinded to reality, crippled in any future effort to form an accurate picture of what was going on with electricity, light, or even matter.

Thus, it is not hard to imagine what the monkey had to

say when the monkey saw the spark produced by a voltaic pile. The voltaic pile, the first storage battery, was described around the same time that Young proved light to be a wave and thus forever removed any attempt to view what was happening with the spark as the movement of measurable electricity seen as light, or the same phenomena rather than two distinct phenomena, one producing heat, the other light.

Connecting alternating elements with different potential differences in a conductive solution produced a spark when the connection was established or broken, just like the spark produced by bringing a rubbed rob close to a source of sparks. We can see the same effect anytime we attempt to put the second cable on the car battery terminal. If we are putting the positive cable on the positive terminal, the spark jumps from the cable to the battery, because, as we shall see, Franklin had it backward, electricity is a negative charge being attracted by a positive pole rather than a positive charge being attracted to a negative pole.

Of course, all we are doing is finding out which way the electricity is flowing and then putting our made-up charges in the appropriate position, but when devotees of the rubbed rod saw the spark on the voltaic pile, they clearly saw opposites attracting. When they discovered that the voltaic pile stored the stuff that produced the charge, the concept of polarity moved into the concept of electricity without a whisper.

The voltaic pile, or any battery for that matter, is made

up of potential differences which are unable to express themselves until they are connected. When they are connected, something flows between the potential differences. If a rubbed rod has a potential difference with respect to the paper it attracts, then the paper will attempt to move toward the rod. If the doorknob has a potential difference with the rod, then a spark will fly to the rod. If two elements have a potential difference, then connecting those elements together will allow some of the stuff that makes the spark from the doorknob flow through the connection from the element with more of the stuff to the element with less of the stuff. If we consider the flow to be positive, then it is flowing to a negative element. If we consider the flow to be negative, then it is flowing to the positive element.

Our concept of opposites attract dictates the charge toward which the flow is moving once we determine what the charge of the stuff that is flowing is.

Opposite charges, however, are unnecessary when we are describing the movement of a charge once different potentials are connected. If two elements have no charge, then each has all of the particles it needs to hold itself together. If we scrape some of those particles off one of the elements, the element then has a charge. The particles that were scrapped off also have a charge. With two like charges exposed, the two want to get back together again.

It doesn't take opposite charges. We aren't dealing with magnets. It merely takes pulling matter apart to expose the charges that are holding the matter together.

When one element has more of a charge than another element, then all that is happening when the two elements are connected is the element with less of a charge will give up charged particles to the element with more of a charge. Like charges are directing the movement of released particles carrying the like charges to areas where there is a greater charge.

The misconception of polarity, placed in the spark by the manufactured concept of fluids carrying the opposite charges dictated by magnets, moved without rational discourse to simple storage batteries because they produced a spark when elements with potential differences were being connected.

The next four or five decades, the first half of the nineteenth century, worked to confuse the matter further while at the same time cementing the concept of polarity in the view of electricity and the matter that produced it. The connector between the potential differences in a voltaic pile became a conductor of the electric fluid and it was rapidly found that bringing a compass needle, a magnet, next to the conductor when it was conducting caused the compass needle to deflect at right angles to the conductor.

In the eighteen twenties, Joseph Henry began to demonstrate that moving a conductor through a magnet's lines of force produced a flow of electricity, and it wasn't long before it was found that moving a conductor at right angles to a flow of electricity also produced a flow of electricity. Wrapping conductors around iron produced magnets when electricity was flowing in the conductor, and thus wrapping conductors around armatures which were set against similarly wrapped rotors produced a mechanical device which could be used to create electricity as a generator or use electricity as a motors.

The key to the process was to produce a circuit in which the electricity could move, and a circuit required that the electricity end up at the point at which it started.

Dauntlessly moving forward with no clear concept of what was going on, we applied polarity to the resulting circuit.

What do we have when we have polarity in a circuit?

We have a contradiction in terms.

When we have a bunch of cars racing around in the Indianapolis 500, we have lines on the pavement which show where the race started and where it is going to stop, but other than that, we just get a little dizzy as we watch the cars go round and round in a circle. In an electric circuit, we have a point at which something is being put into the circuit and we have a point at which something is leaving the circuit, but other than that, we have something that is just going round and round in the circuit. If an opposite charge were causing that something to move, then where in the circuit is the opposite charge? Under Franklin's concept, the spark, now electricity, was a fluid that was positively charged being moved by its attraction to a negative charge. If the point at which the electricity is entering the circuit, the generator, is negatively charged, then why does the electricity move to the load, the light or motor or other item using the electricity? And if the load is negatively charged, then why does the electricity move back to the generator? And if both are negatively charged, why does the electricity move at all?

There is no to or from in a circuit and thus no place to put polarity. But put it we did, and like rubbing out the "north" on the end of a magnet and replacing it with a "south" when we realized that the north pole had to be pointing south if it were to obey the objective laws of physics which require that opposites attract and likes repel, we changed the point toward which electricity flows in a circuit to positive when we realized the electricity wasn't composed of a positive fluid but rather a negative charge even though there is no point in a circuit toward which electricity can flow.

A load is something placed on a flow of electricity that is activated by the flow of electricity. The most common load we have is an electric light bulb, a filament in a vacuum tube. Separating the filament so that it has a gap creates an end from which particles flow and an end into which particles flow. If we put a metal sheath in the tube, we can use it like a gun barrel directing the particles in a stream out of the tube. These streams are called X-rays. These flows were at first thought to be indistinguishable from what came off of the filament as light, but they couldn't be reflected or refracted, they went through some things and were apparently absorbed by other things. Placing magnets next to these flows also deflected the flows.

Which end of the magnet repelled the flow?

The end of the magnet that pointed north, the north end of the magnet.

If the north end of the magnet deflected the particles, then the particles must carry an opposite charge to be deflected.

The north end of the magnet was considered to have a positive charge so that the flow must be made up of negative particles.

The electron was born!

We didn't have the foggiest idea what caused the magnets to attract and repel, we didn't have a glimmer of a notion why one end of a magnet pointed to the geographic north pole, we didn't have the foggiest idea why the magnetic lines of force repelled or attracted the flows in the vacuum tube, but we really didn't need to know squat after we had labeled the flow as flows of electrons. We now knew everything empirical science needs to know about the cause of a physical phenomena. When we monkey saw electricity we could monkey say electron, and when someone asked what an electron was, we could explain that it was a moving charge. It didn't even make any difference in the electron's charge when it was discovered that in accordance with settled empirical law, that opposites attract and that it was the south end of the magnet the electrons were being deflected by, making the electron actually a proton.

When we don't know anything, anything makes sense.

Polarity had been assumed in the battery because the polarity in a magnet required electrostatic attraction to be complimented by electrostatic repulsion. Whether or not electrostatic repulsion existed, it was explained by creating oppositely charged fluids, with the charge in matter being produced by the loss of one of the fluids leaving the charge of the other. Matter, then, contained two charges, a positive and a negative charge, with one of the charges being exposed when the other was removed. If electrons were negatively charged particles, then removing them from matter would leave a positive charged surface toward which the negatively charged electrons would be attracted. Rub the rod with silk and you would scrape off the electrons leaving a positively charged rod. Bring the rod to a source of electrons and, zap, the electrons would jump the gap as an electric spark, replenishing the electrons in the rod and rendering the rod's charge neutral. Just as weight, the assumption that matter has an intrinsic quality that causes it to be heavy, had to be accounted for in explaining how matter was made up, so polarity became a requirement for any explanation of how matter was made up.

Again, we have to pause to understand why. There is nothing observable in physical phenomena that supports the requirement that polarity exists outside of the physical description for the behavior of magnets. Of course, magnets clearly have external flows of something. These flows are actually flowing in a circuit, out the end of the magnet that points north, around in the shortest possible distance between the north and south end given the fact that the flows are moving at the speed of light, and back in the south end of the magnet for their journey through the magnet and out the north end once again.

These flows, called magnetic lines of force, are not hard to see. We can see them in detail by letting them arrange iron filings on a sheet of paper. We can, by having the flows intersect the line of iron filings at right angles, see the actual inductive flows produced by the primary flows externally orbiting the magnetic material. These inductive flows are the same inductive flows we see when we pass a conductor through the iron filings at right angles. When the ends of the magnets out of which these flows are emerging are pushed together, it is easy to see how the flows physically prevent the magnets from coming together, while when the end of a magnet out of which a flow is coming is placed next to an end the flows are entering, it is easy to see why the flows, seeking the shortest distance between two points, will seek to combine the physical material that makes up the magnet.

There's nothing to prevent us from seeing that the magnetic lines of force are really flows of electrons externally orbiting the magnetic material and everything to force us to conclude that they are flows of electrons because moving a conductor through the flows tips the electrons into the conductor to produce flows of electrons in the conductor. There's no reason to believe that rubbing a glass rod with silk does more than rub a few gadzillion of these electrons off of the rod leaving a deficit of electrons toward which electrons will naturally move, but there is nothing to cause us to produce a concept that rubbing the electrons off exposes the charge of an entirely different particle for which we have no evidence, a particle that has its origin in the concept of first a resinous fluid, and then a vitreous fluid both of which had their origin in the ends of the magnet attracting or repelling. In fact, with the movement of magnets the result of external flows, be they flows of electrons or magnetic lines of force, and not opposite charges, there is nothing to make us believe that the attraction of the electrons back to the rod is more than the attraction of electrons to electrons.

The only thing that makes us believe that the removal of electrons exposes a positive charge that then attracts the negatively charged electrons is the concept of polarity we made up to empirically describe the behavior of magnets, a concept that doesn't even apply when carried back to its origin in the behavior of magnets, which clearly have no charge.

And the only reason we have to maintain the belief is that we feel we need an opposite charge to give the electrons a cause for their movement.

However, hidden assumptions forge boldly ahead in the mindless analysis used by empirical science to obscure its ignorance about how matter moves in reality, so that when an atom was constructed to explain how chemical reactions work, and how matter is in turn constructed, the scraped off electrons became a centerpiece of the conceptual atom that resulted. Even though scrapped off electrons would leave a charge that would attract them back if like charges attracted, a deficit of charge carrying electrons meaning that there was a surplus of charge left on the surface which was available to attract the electrons back, the atom had to be constructed with the concept of opposites attract, just as the behavior of matter under light had to operate according to the dictates that white light contained all color and thus matter had to absorb the colors that weren't reflected.

In the beginning decades of the twentieth century, when the concept of the atom was being perfected, polarity was the absolute concept with which all measurements in physical reality had to conform!

Empirical science spent centuries taking something that was apparent, light, the thing that illuminates the world, permeates the universe, controls our existence, and making it nonexistent, an imponderable with no substance between its cause and effect.

It spent the same centuries creating polarity, a term created for convenience to describe the physical behavior of magnets and therefore a symbol, something that was actually nonexistent, and turned it into a prerequisite, a requirement, the dictate that underlies the concept of matter.

Empirical science makes up stuff to explain because it can't explain the stuff that exists.

6. The Electron As The Universal Particle

The formulation of a concept that could explain matter became possible when elements were arranged into the periodic table. The arrangement of elements into the periodic table was possible because elements had two primary characteristics in common. All elements moved in the Earth's field of attraction at a uniform rate, but all moved against the force of the Earth's attraction at a different rate, the resulting force it took to move different elements against the force of attraction being measured on a relative basis by scales and defined as weight. In addition, all elements had a charge which was the element's potential difference with respect to a given environment, and that charge would start at a low point and increase with the weight of the elements until it reached a high point, and then it would start over at a low point once again, giving the table a periodicity when arranged by charge with ascending weight.

The idea was to produce a concept of an atom that would embody everything required to produce the matter we measured in our mind. The resulting concept put a particle in the atom's center, its nucleus, to account for weight. This is really Newton's old corpuscle although nobody has ever found gravity associated with it. The heavier the element, the more of the weight particles the atom had. These particles were named neutrons in deference to their explaining weight without charge.

Next, matter construction required that atoms hold together to each other to form the physical surfaces we see and measure. If we put an electron in orbit around the nucleus, then removing the electron would create a positive charge that would attract other electrons, and if the other electrons were also orbiting atomic nuclei, the atomic nuclei would come along and get together with the original nuclei. The resulting idea for matter formation was that the nuclei would share electrons which would hold the nuclei together into molecules.

A charge had to be created to keep the electron in orbit around the nucleus so a proton was made up and put in the nucleus to balance the charge of the orbiting electron. For every orbiting electron, there was a proton in the nucleus, and when the orbiting electron was lost, the nucleus was left with a positive charge. Thus, the nucleus had neutrons in ascending order to account for weight, electrons in ascending orbits to account for charge, and protons to account for the charge holding the electron in orbit, as well as for some of the weight. Removing or sharing electrons caused atoms to come together which accounted for physical matter.

The concept had a number of drawbacks. For instance, what was causing the electrons to orbit? Assuming that they were held in orbit by the protons, what was keeping them moving. Electricity needs a positive charge toward which to move. What do the orbiting electrons have to keep them orbiting?

We don't have the foggiest idea why planets move, so

what's a little ignorance about a little electron?

How about the mathematical problem of balancing charges. No matter which way you sliced it, exposing a positive charge in the nuclei to attract distant orbiting negative charges, if successful, still left less of a charge than the net positive charge of the two nuclei. Net positive charges repel, so how did this hold the atom together?

The orbiting electrons that were already orbiting the atom would repel the orbiting electrons of the atoms being attracted by the exposed positive charge, so why worry about a little net positive charge. If worse comes to worst, and the net positive charges become too great, as they do inside the conceptual stars created by empirical science, stars which burn from the inside out, we can just say, as does empirical science, that positive charges only act like positive charges, and repel, until they get so close together that like charges begin to attract. If we want empirical proof all we have to do is look at the sun. It's burning without flying apart, isn't it?

Polarity produced a plethora of particles in the original concept of an atom, a plethora which has grown out of all proportions since. The invention of quantum mechanics, with its basic tenet that nothing is predictable, so that all predictions became possible, realized the most successful theory in the history of human cognition. When all predictions are possible, all fantasies become fact, and with awards being given out right and left for these discoveries of scientific fact, discoveries that rivaled the discovery of new continents, new planets, new star systems, it was only natural that the particles that make up the atom also be made up of particles. Finding that the interactions of bits and pieces of matter in cloud chambers left traces in the clouds, and that photographic plates could be substituted for clouds, empiricism began a race to populate the atom with as many particles, anti-particles, colored particles, even personality particles as the awarding committees could stand, requiring an international commission to prevent the proliferation of particles past the ability of all unreality to comprehend.

All of these particles were theoretically identified out of the billions of arcs and swirls found on the photographic plates, and none had relevance to physical reality outside of those arcs and swirls and the theories that identified them as arcs and swirls.

The simple fact is, there is no basis in reality for the proton, there is no basis in reality for the neutron, there is no basis in reality for all of the particles that came after.

The simple fact is that there is no basis in reality for polarity. Polarity is just something we made up and which we use to bend the picture of physical reality, distort it out of all range of reality. There may well be electrons orbiting the nucleus of an atom, but that nucleus does not need an opposite charge to hold those electrons in orbit, it only needs a charge, and there is no writ in reality preventing the particle in the nucleus that accounts for the movement of the nucleus in the Earth's field of attraction that gives the atom its weight from producing the charge that attracts the orbital electrons.

And without polarity, there is no reason that the particle in the nucleus that accounts for weight and which attracts the electrons into orbital clouds around the nucleus can't also be made up of the same electron found in the orbital clouds.

In fact, without polarity, there is no reason why everything, matter, light, electricity, magnetism, heat and combustion cannot be explained by a single charged particle.

Why does empirical science conclude that electrons don't make up light? Why does empirical science deny that what we pay for our electric bills is not what is coming out of our light bulbs.

There are several reasons for the scientific fact that light is not made up of electrons.

The conceptual creation of polarity, with its dictate that positive charges exist to produce movement in negatively charged electrons blocks rational analysis of an elementary particle that could account for both light and electricity because light has no apparent charge and doesn't in any event need a charge to explain its movement because it is nonexistent between its cause and effect, a ripple in the universal pond.

In addition, because an electron is part of matter, it must have mass, although in reality mass itself is merely a concept derived to describe what has to exist when the inverse square law dealing with falling objects on the surface of the Earth is misapplied to motion in space. It's a scientific fact and therefore incontrovertible that, as we have seen in Chapter Four, light has no mass, otherwise matter would steadily lose, or perhaps gain, weight as it burned.

We have, when it comes to attempting to understand what light is, how it is produced and where it goes, slowly and silently, over the years, tied ourselves incrementally into a pile of knots so tangled that we are incapable of even seeing the obvious realities around us. Thus, when Einstein demonstrated the photoelectric effect in which light falling on matter released electrons, instead of opening up the old debate about whether light was a wave or a particle settled inadequately by the water wave analogy in Young's two-slit experiment, empirical science shrugged its shoulders and forged boldly ahead, saying, golly gosh, light measures as a wave and light measures as a particle, so light must be both a wave and a particle. Pity the poor idiot that questioned how such a thing could be pictured. We could walk through a laboratory door activated by a photoelectric cell and perform Young's experiment to see that light was actually a wave/particle or a particle/wave.

Forget the fact that we had no idea what was happening in Young's two-slit experiment, and forget the fact that we had no idea how it was that light released electricity in a photoelectric cell, we could monkey say what we monkey saw, interference patterns and the photoelectric effect, so that words could become comprehension. What more could we want out of science?

The fact that light was both a wave and a particle was a reality because it could be scientifically shown on a repeated basis in the laboratory to be fact.

The production of a stream of something in a cathode ray tube that led to the visualization of a charged particle known as the electron led to the discovery of electrons everywhere in the environment. The production of light in a vacuum tube was not seen as analogous to the production of electrons in a cathode ray tube, and even today, with our entire civilization based on the production of light from the streams of electrons fired from cathode ray tubes striking dots on a television screen we still visualize light as made up of something other than electrons, calling what makes up light photons, energy quanta, just as we eventually organized all of the electrons we found in the environment into other particles, things like thermions, for instance, where electrical effects were associated with heat.

We even "see" the stars by converting their light to measurable electrical flows.

Today, in our electronically driven world, light is interchangeable with electricity, with light emitting diodes converting electricity into light and photodiodes and charge coupled devices reversing the process. The most obvious conversion of electricity into light goes entirely unnoticed, the steady hum of the electric meter as it spins out the consumption of electricity by the lights burning brightly in the night. If we pick up a book on electricity, we are told that electricity is a moving charge and then we are dropped into an endless series of mindless mathematical formulae that describe what happens with electricity under all conceivable circumstances, the majority of which we fabricate in our minds. If we pick up a book on light, we won't be holding a book on light, we will be holding a book on wave theory and we will be dumped into an esoteric world of vectors and wave theory equations that have little or nothing to do with light itself.

How can we get to the root of what's going on in physical reality without some sort of concept of what it is that is underlying physical reality? Empirical science has constructed an atom that gives us a hazy idea of what's going on in chemical interactions, and an internally contradictory picture of what's holding matter together, atoms losing negative charges to expose positive charges which attract negative charges through the remaining negative charges which don't repel the attracted negative charge and which end up with the net positive charges dragged close by the attracted negative charge so that the positive charges now attract rather than repel.

What we need is an elementary particle that we can use to construct an atom that actually describes matter in physical reality and in addition describes how that matter interacts with the hidden forces in physical reality.

To do this, the particle also has to describe those hidden forces!

Today we have particles to describe different aspects of the matter that makes up physical reality and more particles to describe the hidden forces with which that matter interacts. We have photons, electrons, protons, neutrons, thermions, just for starters, and if I wanted to fill up this page and the next, I would have but to buy any book purporting to explain what's going on in physical reality for an endless list. Any time we think we find something new, we say, well, what are we going to name this one, naming things being the end goal of the empirical process because it is only by putting words into recall instead of actual pictures that explain physical reality that we can fool our lazy minds into thinking we actually know something. When it comes to naming forces, we have electricity, a moving charge, magnetism, molecular magnets, the strong force and the weak force, and gravity, embedded in matter as mass but not in the atom that makes the matter up.

What we need to do in place of mindlessly naming things, the empirical process of monkey see, monkey say, is to sit down and figure out what is at the basis of what we measure. We have measured enough, we have measured too much, we are measuring things that don't exist and using them to create other things which we promptly go out and spend billions of dollars trying to measure.

It's time we figured out what's going on. And to figure out what's going on, we have to synthesize an elementary particle with properties that describe what's going on, rather than mindlessly creating particles for things we fantasize are going on.

How do we create a particle with properties that describe what's going on?

Creating the particle is simple. We do it all the time. Particles are conceptual things that we make up to explain what exists. Thus, all we have to do is realize that light, electricity and the atoms that make up physical reality are in fact made up of an elementary particle, the same elementary particle, and we are halfway to our goal.

We have to start with the statement that there is matter and there is not matter, and that matter occupies nothingness, the nothingness being the "not matter" portion of the two part statement. Once we have that firmly in mind, then all we have to do is say that anything that exists in nonexistence is matter and is made up of the same elementary particle.

We may see matter as a solid, we may see it as light, we may see it in nature as minor electric flows, or as magnetic flows, or we may produce it in chemical interactions, or we may create gigantic electric flows, but no matter what we are dealing with, we are dealing with something, with matter in nonexistence, and with matter that, regardless of how it presents itself, is nonetheless made up of the same particle.

Thus, light will become a flow of these particles, matter will become conglomerations of these particles, electricity will be a flow of these particles, and when something moves in physical reality, Sir Francis Bacon's long ignored secret motion of things, then it will be the result of something physical happening with the particles, some interaction of the particles one with the other, the particles being all the same particle.

We are not dealing with occult forces such as gravity, and its endless search for the graviton, or magnetism, and its mindless molecular magnets lining up all in a row, or a moving charge, whatever that is, or some sort of absurd strong force that mysteriously overcomes the repellent force of the imaginary positively charged protons in the nuclei of atoms so that atoms can combine in the process of fusion.

We are not dealing with waves through a medium or atomic vibrations of orbiting electrons somehow mysteriously otherwise immobilized in orderly orbits, a belief that we have to accept on faith because otherwise the opposite charges we made up and placed in the atom would destroy the atom.

We are talking about forces in the here and now, current forces that exist as a part of reality and we are talking about a single elementary particle with properties that cause the matter it makes up to interact in a way that the physical reality we see and measure results.

If we have a single elementary particle at the basis of measurable observation, then how do we derive its properties?

The way we find out about anything in reality, by simple observation!

What are the two observations concerning physical reality that are obvious, pervasive, and rock solid unquestionably real?

One is physical reality.

While there are those that find it intellectually stimulating to deny the existence of physical reality, it is physical reality that we are attempting to deal with, and denying its existence will end the inquiry. Physical reality exists, and for the most part it is solid enough to do away with us if we collide with it with sufficient force. This means that physical reality is something that is solid, has material substance, that we can knock our heads against. Any particle that makes up this physical reality must be held together by some sort of force.

We are not going to be able to get away from an attractive force, but we don't need a bunch of attractive forces, the attractive force of a moving charge, the attractive force of a proton, the attractive force of two poles of a magnet, and the attractive force of gravity causing objects to drop, to name four. All we need is an attractive force in the particle that makes up matter and which holds the matter together so that we see the solid physical matter that is us and which we can physically move.

If we have matter, then we have something holding the particles that make up that matter together, and that force can be a property of the particle that makes up the matter.

As soon as we propose a force that must exist in the particles of matter in order for that matter to be held together into solid physical reality, we have to describe how that force acts.

We are mentally crippled by an empirical science which demands that we only accord reality to that which we can physically measure and then goes on to create an occult force, gravity, that diminishes when it is not used, and more to the point, doesn't diminish when it is used. Gravity is conceptualized as decreasing inversely with the square of the distance over which it acts, but when it acts, it is never used up. It can attract the moisture from your nose so that it dribbles down your lip, move on without having been diminished one whit by the effort, paste your hair in place, proceed through the atmosphere diminished uniformly by distance but not by the force it takes to hold the molecules of air against the surface of the Earth, work against the jets of a passing plane with ease, go on to keep the moon effortlessly in orbit, affect the straight-line motion of Jupiter, and cause difficulty in the sinus cavity of some unknown creature inhabiting some obscure planet at the end of the universe and the beginning of time.

We would be better off consulting an astrologer to divine the meaning of the universe than attempt to use such an absurd force to explain physical reality. The force of attraction that we need to use in describing how matter is made up has to be a force of attraction that is used up holding the matter together. Otherwise we would be crippling our efforts before we started.

Just how do we visualize a force that is a property of a universal particle that comes together to present the matter we measure, physical reality? How do we visualize a particle using up a force?

One way to visualize it physically would be to take little magnets and add them together. Each magnet has the same amount of magnetic force, and if we put two of them together, we would find that it was pretty hard to pull them apart. However, if we added a third, then the total amount of force holding the three together would be increased, but the force holding each together would be reduced. It is easier to pull two magnets apart when they are part of three magnets held together than when they are just held together by each other.

If we add a fourth magnet, we bring more magnetic force to the structure, but it becomes still easier to pull the individual magnets apart. We can continue adding magnets to the structure until the magnetic force between the individual magnets is so weak that the magnets can barely hold another magnet to the structure.

This concept of net force diminishing as additional force is added due to the requirement that the amount of force required to hold the structure together increases faster than the additional force added has to be inherent in the force of attraction that is the property of the particle that makes up matter. Otherwise, there would be no limit to the size that matter could achieve.

One particle has to have an attractive force. Two particles each have to have the same attractive force. The two particles are attracted to each other. When they come together, the force of attraction holds them together, but in doing so, is used up to some extent. The two particles have double the attractive force of one particle, but a little bit of the attractive force of each is used up so that the attractive force of the resulting structure is less than twice the attractive force of the two particles if they were not together.

This process continues.

A third particle is being attracted to the two that are already held together, but the resulting force of the three combined particles available to attract additional particles is less than three times the force of the three particles individually.

We can visualize the process as one in which the property of attraction of each particle is being used up to hold the overall structure together. As each additional particle is attracted to the structure, it adds its attractive force to the overall structure, but the total attractive force of the overall structure becomes progressively less until it reaches a point at which all of the attractive force of the particles are used up in holding the structure together and no new particles can become a part of the structure.

The structure thus formed would be the basic unit of matter. It would be a unit of matter which would always be identical. The total number of particles that could come together as a result of each of the particle's attractive force would always be the same because the amount of that attractive force is always the same. Because all of the basic units of matter formed out of these individual particles would contain the same number of particles, they would all be the same size.

In other books in The Copernican Series, primarily *Atoms, Stars and Minds*, the third volume, I discuss how big this elementary particle is and what the force of attraction is. Basically, the particle is merely big enough to define nonexistence, and the attractive force is not really so much an attractive force, as an affinity propensity, the propensity of each of the particles to occupy the space of all other particles.

Because affinity propensity is basic to an understanding of how light is produced, we have to visualize how affinity propensity works.

Affinity propensities work, they attract other affinity propensities, when there is an excess of affinity propensity, and excess affinity propensities disappear when they are satisfied, when they have attracted other excess affinity propensities.

One particle standing alone has all of the affinity propensity that it can have so that all of its affinity propensity is excess affinity propensity, affinity propensity available to attract additional particles.

By eliminating the concept of positive and negative charges in favor of a charge that only attracts like charges, we have merely simplified the conceptual situation that had electrons orbiting a nucleus with an equal positive charge. When we took a few electrons away from a positively charged nucleus, we were left with a positive charge.

Without the concept of positive and negative, we don't have a positive charge when we remove electrons, we merely have an excess charge, an affinity propensity which is now exposed and capable of attracting additional affinity propensities. When the first particle, and we can call them electrons because the affinity propensity that they evidence is the same charge that the electron is measured to have, when the first electron is attracted to another electron, each has its affinity propensity in full excess. All of the affinity propensity in each electron is available for use in holding the two electrons together.

When the two electrons come together, the amount of affinity propensity left over after the affinity propensity of each electron is used to hold the electrons together is excess affinity propensity available to attract other electrons. As the electrons are attracted to and become a part of the forming unit, the amount of affinity propensity in the overall unit is increasing but, because the amount of affinity propensity needed to hold the overall unit together is also increasing more rapidly, the excess affinity propensity available to attract additional electrons is decreasing at an increasing rate so that eventually, when some gadzillion or so electrons have come together into a single basic unit of matter, all of the excess affinity propensity of the unit is being used up holding the individual electrons together into the unit.

If we should all of a sudden do something to remove

some of the electrons, then we would have a unit with an excess affinity propensity, a unit with a charge, a unit which would appear in our detectors differently than a unit that had no excess affinity propensity.

The concept of excess affinity propensity drives the universe. We will see it in magnets, light, electricity, chemical interactions, and in all of its glory in the mechanism that regulates the uniform expansion of electromagnetic emissions, the mechanism whose by-product produces the movement of matter we attribute to gravity.

There would, however, be little need to hold the electrons together unless there was a force that was attempting to keep them from staying together. Without such a force, there would be no need for an affinity propensity. The electrons would just come together by chance, and if there were an affinity propensity, then all of the electrons would come together in a big clump. The universe would simply be one big pile of electrons in the emptiness of nothing with nothing going on anywhere.

For the affinity propensity to be used up, and thus for there to be an excess affinity propensity available for use in attracting additional electrons, there has to be a property of the electrons that is always attempting to keep them from forming into these units of matter, the force which actually limits the number of electrons that can come together into the basic unit of matter.

Where can we look in reality to find a force that could

exist in the electron, along with the electron's property of affinity propensity, which would limit the size of the basic unit of matter by having to be overcome by the affinity propensity?

We don't have far to look. All we have to do is look at electricity, the movement of electrons. Electricity moves close to the speed of light, the reduction in its speed from the speed of light being attributable to the resistance of the conductor through which it moves.

But we have an even more revealing example of the second property of the electron, the particle we are going to use to explain light. If we take light and pass it through water or glass or some other conductor, we slow it down. When it re-emerges from the conductor, it speeds right back up.

There can be no clearer example, nor better demonstration of the fact that one of the properties of the electron, the basic particle that makes up matter, is motion, the ability to travel at rest at the speed of light.

What does at rest motion mean?

The best example of at rest motion is probably a rock at rest on the ground. The rock looks like it is perfectly still and yet we know that it is traveling close to a thousand miles per hour as the Earth spins on its axis, about sixty-seven thousand miles per hour as the Earth orbits the sun and about three hundred and twelve miles per hour as the sun moves slowly around the rim of the galaxy of which it is a part. Contrary to what Newton would have us believe, matter comes to rest with respect to the forces acting upon it, so that even though the rock is moving in three different directions at once, all of its movement is at rest movement. If the Earth, which is at rest with respect to the force acting upon it all of a sudden came into contact with another planet moving in the opposite direction, the resulting bang would be considerable. The motionless rock on the Earth's surface would probably end up in smithereens.

At rest motion doesn't mean that something is not in motion. It means that it is in motion while it is at rest with respect to the forces acting upon it.

The solar system is an outstanding example of matter whose motion is at rest motion, planets at rest with the forces acting upon them moving with respect to one another.

The property of at rest motion in the electron means that the particle is at rest with its own inherent force when it is moving at the speed of light, one hundred and eighty-six thousand miles per second. If we slow the particle down some way, then we have restrained its at rest motion.

When we restrain an electron's at rest motion, we are increasing its potential for motion. When light enters a medium such as a prism, the at rest motion of the electrons slow down and their potential energy increases, the potential energy being the potential of the electrons to regain their full at rest motion. When the electrons re-emerge from the prism, they regain their original speed, but do so with changed frequencies, the lower frequencies of red and green light breaking out on either side of the highest frequency, what we see as yellow light.

When the basic units of matter are forming, the affinity propensities of the individual electrons are overcoming their at rest motion so that when the basic unit of matter is formed, all of its at rest motion is potential, each electron in the basic unit of matter has the potential of regaining its full at rest motion. When the electrons are moving at the speed of light, either as light or as electricity, then the at rest motion has to a great extent overcome the affinity propensities of the individual electrons.

It doesn't take too much imagination to see Einstein's great conceptual equation equating energy with matter at work here. Matter, made up of basic units containing electrons whose affinity propensities have overcome their at rest motion, has a potential energy equal to the energy that would be obtained if each of the electrons making up the basic units of matter were to regain its at rest motion at the speed of light. The injection of the speed of light into the conceptual equation merely allows the mind to grasp the magnitude of the amount of excess affinity propensity, and thus energy, locked up into the basic units of matter. Using a number like "one" for the speed of light as we do with astronomical units for spatial distances would, of course. have defeated the conceptual impression the equation produces.

What we have, then, is an elementary particle, an

electron, with two properties, affinity propensity and at rest motion. When the property of affinity propensity has overcome the property of at rest motion, then we have physical matter. When the property of at rest motion has overcome the property of affinity propensity, then we have energy, light, the expanding electromagnetic spectrum.

We will see that combustion is simply the process of the electron's property of at rest motion overcoming its affinity propensity so that it flies away from the point of release at its at rest speed of light.

However, before that can be visualized with clarity, we have to see how the property of affinity propensity overcomes the property of at rest motion to form the atoms that make up matter.

7. Constructing An Atom That Describes Reality

The concept of polarity was carried into the concept of electricity in order to explain why electricity moved. Light, which didn't exist, could travel from one place to the other without any necessity for a motive power because it was merely a disturbance in some sort of medium, still widely thought of as the aether. Although Maxwell's equations provide a hazy concept of a motive power for light in the interaction of electric and magnetic fields, that concept is immediately vitiated by its three dimensional representations which have arrows going every which way. Nothing needed to be said about the motion of water waves because the motion was right in front of our eyes. We monkey see the waves and monkey say the water, and the motion is clearly understood.

Water waves move on their own, so why shouldn't light?

Electricity on the other hand had to be thought of as moving from one place to another because if we create a gap between the poles of a battery, we get a spark. A spark was like lightning, and lightning clearly moved from one place to another.

We needed something to monkey say about its movement so that we wouldn't be puzzled when we monkey saw it.

We concluded that the negative charge of the spark

moved because a positive charge was attracting it.

When a vacuum is produced, the affinity propensities associated with the atoms and molecules of atoms that would have occupied the area of the vacuum are eliminated. By eliminating excess affinity propensities inside the vacuum tube, the vacuum tube creates excess affinity propensities where there are still atoms and molecules of atoms outside the vacuum. When an electric current is passed through the vacuum that contains a gap in an electrical circuit, a stream of electrons will flow across the gap. If a metal sheath is used to direct the electron flow, the electron flow will be attracted through the tube to the excess affinity propensities at the boundary of the vacuum.

This is the cathode ray tube used to categorize the flows as made up of electrons and conclude that electrons weren't the same thing as light, that they weren't the light that occurred when the gap was connected with a load.

Of course, the circuit out of which the electrons were drawn has no negative or positive poles like a battery, but this problem is easily overcome by inserting a couple of "p"s and "n"s or "+"es and "-"es in the circuit to keep the viewer oriented as to which way the electricity is flowing in the circuit, even though we are only talking about direction of flow and therefore a simple arrow would do the trick. A simple arrow, however, would raise the question the polarity symbols answer by implication. Without the little polarity symbols the inquiring mind might ask what is making the charges move. Because we are dealing with settled areas of scientific fact, questions such as this are not relevant because the little symbols provide all the answer that anyone could ever need in an empirical world. Monkey sees the "p", monkey says the "p" is attracting the electrons just like one magnet attracts another.

With at rest motion being a property of the electron, the question is not what is causing the electrons to move through the circuit, but rather what is keeping them from moving at their full at rest speed, the speed of light.

And the question of where electrons move is not a question of polarity, a negative charge needing a positive charge to cause the negative charges to move, but rather a question of excess affinity propensity. An excess affinity propensity is an area in which there is a deficit of electrons with respect to a particular environment so that the area will attract additional electrons to bring it into balance with that environment.

To understand how and why electrons flow, we need a clear picture of the matter that generates the flows. To obtain a clear picture of the matter that generates the flow, we have to have a workable atom that would account for both the matter and the flows. Because the atom we have conceptualized doesn't explain physical reality, we have to take the particle we have hypothesized, the electron, with its at rest motion and affinity propensity, and use it to construct an atom that actually explains what is going on in the physical reality of which we are a part.

The classical concept of the atom, the concept that created a successful chemistry based on the twin concepts of valence theory seated in atomic structure and chemical bonding based on shared electrons, has a nucleus containing neutrons to account for weight and protons to balance the charge of the electrons orbiting the nucleus and also to account for weight. The electrons orbiting the nucleus in turn were (and in many cases still are) placed in shells to account for the periodicity of the elements, basically their weight arranged to the extent possible by their electrical characteristics. This classical concept of the atom has devolved, strangely enough as a result of the measurement of light as a particle and the failure to shake the concept light as a wave produced by incessant viewing of Young's two-slit experiment, into a fuzzy picture of the atom as a nucleus of neutrons and protons made up of forty or fifty other particles surrounded by a cloud of orbiting electrons.

The neutron, a concept created to account for weight, is unnecessary. Weight is a concept that has been totally debased by the concept of mass, although to be truthful, what mass replaced, *gravitas*, wasn't much clearer. What gives something weight has never been able to be separated from the thing that weighs, and thus we accept gravity as a force that gives things weight because gravity is a part of the thing itself.

Weight is the result of the relative force it takes to stop atoms from moving in a field of attraction. This is what the surface of the Earth does. It stops atoms from moving in a field of attraction.

Objects fall at a uniform rate regardless of the nature of the atom that makes them up because there is a force that is acting on individual units in the nucleus of each atom. Thus, an atom with a hundred units in the nucleus will fall at the same rate as an atom with fifty units in the nucleus because what is making them fall is acting on the individual units that make up the nucleus and not on the nucleus. When the two atoms fall, one with a hundred units, and one with fifty units, weight is not a factor in the rate that they fall.

It is only when an attempt is made to stop the acceleration that results from atoms falling in a field of force that weight is created in atoms. If we place a physical plane, such as the surface of the Earth, between the force causing atoms to accelerate and the atoms themselves, so that the atoms come to rest on the plane, so that they cease to accelerate, then the force with which the atoms press against the plane will be determined by the number of units tied together in the nucleus of the atom.

While the force acts on individual units in the nucleus, when the individual units in the nucleus are tied together with a force, the force required to move any single unit becomes the force that it takes to move all of the units tied together in the nucleus. Thus, if we wanted to lift the atom with fifty units tied together in the nucleus, and we wanted to lift the atom with a hundred units tied together in the nucleus, we would need to exert twice the force to lift the hundred unit atom than we would need to exert to lift the fifty unit atom.

It is not the units that have weight, it is the force that it takes to move an atom against a force of attraction that we perceive as weight. And it isn't the increasing number of units that increases the weight, it is the increasing number of units tied together in the nucleus that have to be stopped or moved simultaneously against the force that we use to categorize atoms as having different weights.

Atoms don't really have different weights, because weight is an expression of the force it takes to keep something from moving in a field of attraction. Atoms simply have more units tied together in their nuclei, and the more units tied together in the nucleus, the more force it takes to keep the atom from moving in a field of attraction.

It is this increasing force that we perceive as the increasing weight of the atoms that make up the elements.

If we have a basic unit of matter held together in the nucleus of an atom, then the basic unit of matter is all that is required to explain the result we get when we put different atoms on a balance or calibrated spring and get different effects, the progressive effect of elements in a field of attraction that we call weight. The attractive force acts individually on the basic unit of matter so that if the matter is not restrained, all matter accelerates at a uniform rate, but if it is restrained, then the effect of the attractive force is felt on the total number of basic units tied together in the nucleus.

Atoms, Stars and Minds, goes into detail about how matter forms in the absence of a field. Basically, individual electrons come together into the basic unit of matter that makes up the nucleus of an atom. This is not some long drawn out process, but rather an instantaneous result of the elimination of a field that would otherwise attract the excess affinity propensities of the electrons. With only the excess affinity propensities of other electrons to be attracted to, the electrons begin to come together into the basic units of matter. Two electrons together have a greater excess affinity propensity than one, and as soon as two electrons come together, the greater excess affinity propensity of the two becomes a focal point for the movement of other electrons in the area.

As a basic unit begins to form, its excess affinity propensity attracts ambient electrons, electrons which aren't attached to another excess affinity propensity, to the structure. If there are sufficient ambient electrons in the area, then the basic unit forms instantly. The reason that we don't measure matter formation in the physical reality in which we find ourselves is because we live in a world of excess affinity propensities in which ambient electrons move from one excess affinity propensity to another and thus never get the chance to come together into the basic units of matter that make up our reality.

The surface of the forming unit is spherical because that is the way matter forms, in all directions from the center, the result being the highest volume with the lowest surface area that matter can occupy.

As the electrons attach themselves to the forming basic unit of matter, the unit begins to produce a physical area and the excess affinity propensity begins to reduce as this area expands. As the excess affinity propensity of the forming structure is reduced, it eventually reaches a point at which it cannot attract another electron to become a part of its surface, and the basic unit of matter reaches its full size.

Because every electron that makes up this basic unit of matter has the same amount of affinity propensity overcoming the same amount of at rest motion, one basic unit always contains the same number of electrons as any other basic unit. It may be more than a gadzillion or it may be less, but whatever it is, it is always exactly the same number.

Even when the last electron has become a part of a formed basic unit, and the excess affinity propensity of the unit can no longer overcome the at rest motion of a single electron, the unit will still have an excess affinity propensity. It may not be enough excess affinity propensity to attract an additional electron and overcome its at rest motion to become a part of the unit, but it is still an excess affinity propensity and that excess affinity propensity will attract electrons. The electrons it attracts will have their at rest motion overcome to the extent that they begin to orbit the structure.

This would result, if there were no other basic units forming in the area, in what we visualize as a hydrogen atom, an atom with the fewest basic units in the nucleus and a cloud of electrons orbiting that nucleus. We see these atoms everywhere because any unit that breaks away from a molecular structure will have an excess affinity propensity that will attempt to attract an orbital cloud of electrons. This is why we measure the proton we made up when we knock basic units off the surface of matter as short lived alpha particles. These particles register as a charge in our detectors until they obtain an orbital cloud which balance their affinity propensities at which point they disappear from the detectors designed to detect only charges.

We don't need the proton in the nucleus of the atom, either. Polarity is not necessary because there is no reason to apply the concept that opposites attract, made up to describe the physical behavior of magnetic material, to an atom that can be constructed out of an electron with its two opposing properties of at rest motion and affinity propensity. The opposing forces implied by the concept of polarity are, in effect, the electron's opposing forces of attraction and motion. All we need is a basic unit composed of electrons which has, once it has overcome its last electron's at rest motion, an excess affinity propensity which can attract a cloud of orbiting electrons.

Thus, we are left with an atom, albeit a simple hydrogen atom, constructed solely out of the elementary particle, the electron, with its two properties of at rest motion and affinity propensity.

Matter formation occurs in the absence of a field simply because a field itself, a flow of the same electrons, has an excess affinity propensity which would reduce the availability of ambient electrons needed to form into the basic unit of matter. This runs us straight into another firmly held scientific fact with no basis in reality, the belief that all atoms start out as hydrogen atoms, and as stars burn, these hydrogen atoms form into all of the heavier atoms in the universe. This visualization follows along with the notion that the sun burns from the inside out when everything in our experience burns from the outside in. Iron is supposed to form somehow out of gravitational forces at work in the interstellar clouds of gas as they swirl like a pizza maker's creation in empty space, the scientifically based view that underlies all of the poetry dealing with our having as our source starlight.

In reality, when we stick something in a fire, it breaks down as far as it can in the given heat, and if we make the fire hotter, it breaks down even further, and if we make the fire as hot as the sun, then it breaks down to what we measure on the surface of the sun, hydrogen atoms. Under the monkey see, monkey say process of empirical science, we monkey see hydrogen on the surface of the sun, we monkey say the sun is made out of hydrogen through and through when in reality the hydrogen we measure on the surface of the sun is left over after the atoms with the highest number of basic units in their nuclei have broken down into atoms with less basic units in the nuclei until everything is broken down into atoms with only a single basic unit in their nuclei, and then even that is broken down.

The reason that stars, and the center of the Earth are made out of atoms with the highest number of basic units in the nuclei that can come together is because that is how the atoms form in the absence of a field. When the basic unit has attracted the last electron to its structure, it can attract orbiting electrons if there are no greater excess affinity propensities around to attract them away. Because matter formation is going on, because basic units are forming, any basic unit that is in the process of forming would have a greater excess affinity propensity than a formed basic unit. There are therefore always greater excess affinity propensities around during matter formation to keep electrons from orbiting the basic units as they form.

With an excess affinity propensity unable to attract orbiting electrons, the only thing available for a basic unit with excess affinity propensity to attract is the excess affinity propensities of the other formed basic units.

Thus, just as the electrons formed into the basic unit as a result of their excess affinity propensities, the basic units come together as a result of their excess affinity propensities until all of the basic units that can be held together by their excess affinity propensities have come together.

Because all basic units are identical, the resulting nuclei that form all have an identical number of basic units and it is these nuclei that begin to grow out in a sphere to form the matter, the stars and planets that we measure to exist in the universe.

It is not until combustion begins to occur on the surface of matter formation that this atom, with the highest number of units in its nucleus, begins to break down into atoms with fewer units in their nuclei, and it is not until combustion occurs that there are enough electrons in the ambient field to be attracted into orbital clouds around the newly created elements and compounds of elements that result from combustion breaking down the atom with the highest number of units in its nucleus.

The result of combustion is a universe of physical matter in which the nuclei, attracted to one another by their excess affinity propensities, are physically kept apart by their clouds of orbiting electrons. The nature of this binding is set forth to some extent using hydrogen and oxygen atoms in the evaporation and condensation of water and the formation of ice from water and water from ice in the fifth volume of the Copernican Series, *How the Weather Really Works!* Basically atoms reach a balance with each other which is dictated by the mutual attraction of their nuclei with the physical opposition of their orbiting clouds of electrons. Knock electrons out of the orbiting clouds and the resulting increase of the excess affinity propensities of the nuclei causes them to move closer together. Moving closer together causes the orbiting clouds to interfere which moves the nuclei back apart.

Balance is reached when the attraction of the nuclei and the physical opposition of the orbital clouds are in equilibrium.

To get a visualization of what is going on inside the

Earth, we can picture the Earth with a surface similar to the surface of the sun. As with anything that is surrounded by an area with less activity, less heat, the boundary area of cold, empty space begins to cool the Earth's emitting surface so that the complex elements underlying the combustion begin to break down less and less. A crust begins to develop between the receding sphere of combustion and cold empty space. Beneath this crust, perhaps eighty to a hundred miles today, there is still the receding surface of these atoms undergoing combustion, the process of converting the atoms with the highest number of units in their nuclei down into atoms with less units in their nuclei. Beneath this receding sphere of combustion, however, the original closely packed nuclei with the highest number of units still exist.

Our infinite variety of elements and compounds of elements comes from atoms breaking down as they undergo conflagration, and mixing in the conflagration as they do so. The receding ball of fire beneath the surface of the Earth that is the boundary in which the conversion is taking place is in turn producing an emission field of the highest frequency which, as it moves through the intervening layers of crust, has its frequency reduced until it emerges as very low frequency emissions at the surface of the Earth, emissions which presently constitute approximately three percent of the Earth's heat as measurable by anyone with a radio dish.

Emissions are flows of electrons which make up a field, and because flows of electrons are made up of electrons that have at rest motion and affinity propensity, the excess affinity propensity of a flow of electrons such as are emitted at the surface of the Earth as low frequency radiation can replace the orbital clouds of electrons that satisfy the excess affinity propensities of the nuclei of atoms.

In the atomistic scheme of things, in which everything is thought to occur locally within the atom, the effects that extend beyond the boundaries of the atoms, or the particles devised to make them up, are considered to be a field. Thus, affinity propensity, which can be thought of as an area around the electron that attracts other electrons, can be considered a field.

In the overall scheme of things, an expanding sphere of electromagnetic emissions either as a result of combustion or of fabricated electromagnetic oscillations, is also a field albeit a field that has its affinity propensities balanced.

Field replacement deals with the substitution of one field for another field. One field can be substituted for another field if one field is more stationary than the other field.

The substitution of a more stationary affinity propensity for a less stationary affinity propensity is what produces ourselves and all of the activities we measure in physical reality.

The best example of a nonstationary affinity propensity is an ambient electron. Ambient electrons fill the clouds in the sky between the time that the ice flecs that make them up condense into water molecules and the lightning strikes. They hug the ground on a cold morning when the sun strikes the surface of the Earth, or later in the day, asphalt, causing mirages that act like mirrors.

Ambient electrons are electrons with no purpose. They are not attached to any particular excess affinity propensity, and thus their affinity propensities have not been used up. They move randomly as a result of their at rest motion, colliding with one another, and traversing the open spaces between the atoms that make up the molecules of air. They are everywhere and are evidenced as a component of temperature along with the directed emission fields that result from combustion, the emissions from the sun and the low frequency emissions from beneath the surface of the Earth.

A more stationary affinity propensity is the electron that makes up the clouds that orbit the nuclei of atoms. These electrons have some of their affinity propensity used up satisfying the excess affinity propensity of the nucleus of the atom. The amount of affinity propensity used up is equal to the amount of at rest motion overcome in the process of orbiting. The characteristic of an orbiting electron is that its hold on the nucleus is tenuous indeed because, balanced by its at rest motion, the slightest disturbance will send it on its way at its at rest speed of light.

The boundary surface of matter is a fertile area of interchange for these orbiting electrons. When the silk handkerchief slides down the length of the rod, it is actually scraping electrons out of the orbits of the surface atoms that exist at the boundary of the rod with the air. The atoms at the surface of the rod are left with a deficit of electrons in their orbiting clouds which means that their nuclei have an excess affinity propensity. This excess affinity propensity is evidenced by its attraction of the electrons orbiting the boundary atoms of the bits and pieces of paper, and the physical matter of the paper follows the orbiting electrons which are held to the nuclei of the atoms that make up the paper too tightly for the rod to attract them away.

Orbiting electrons are more stationary than ambient electrons but less stationary than the electrons that make up an emission field. At first blush, this would seem to be contradictory, because the electrons in an emission field, for instance an expanding sphere of light, are traveling at one hundred and eighty-six thousand miles per second. It is hard to think of anything less stationary than something traveling that fast. However, we are looking at stationary affinity propensities, and not stationary electrons. The stationary affinity propensity in a flow of electrons, be it a flow of electricity or a flow of light, is the result of presence.

Presence is a simple concept used to describe the fact that at any point in a flow of electrons there is an electron. If we position ourselves at a point in a flow of electrons, then at that point there will always be an electron because one has just vacated the point, another is occupying the point, and a third is just about to enter the point. When we sit and watch a flowing stream of water, the only way we know that it is flowing is if there is a leaf or bit of foam being transported by the surface. As far as we are concerned, one molecule of water is the same as another molecule of water. There is always a molecule of water at any point on the surface of a flow of water. It is not the same molecule, but it is a molecule of water nonetheless and has all of the characteristics of a molecule of water so that at any time at any point on the surface of a flow of water there is a molecule of water.

At any point in a flow of electrons there is always an electron, so that at any point in a flow of electrons there is the effect of an electron. The at rest motion of an electron in a flow of electrons is used moving the electron forward to form the moving stream of electrons, and some of the affinity propensity is used up holding the flowing stream of electrons together. However, at any point in the stream of electrons there is an excess affinity propensity that results from the presence of an electron at that point at any one moment as long as the stream of electrons is flowing.

How does this point affect the ambient electrons in the area?

It attracts them.

How do ambient electrons respond when they are attracted to the affinity propensity at a point in a flow of electrons?

They orbit the flow.

How do electrons orbit the flow?

They do so at right angles to the flow.

Electrons attracted to the excess affinity propensity of the nucleus of an atom orbit in a cloud because the nucleus of an atom is a point in space and thus the orbits that can be established around the point are any orbit on the surface of the spheres that surround the point.

However, the electrons that orbit a flow of electrons cannot orbit in a cloud because they would interfere with each other and knock each other out of orbit around the flow. The only way that an electron can orbit a flow is if it orbits the flow at a right angle to the flow. There are electrons orbiting the contiguous points to each point in the flow so that for any electron to orbit without knocking out the electrons next to it or without being knocked out by the electrons next to it, the electron has to orbit on a plane at a right angle to the flow.

This is the source of induction in electricity!

Induction is the flow of electrons around a conductor containing a flow of electrons. When conductors with flows moving in the same direction are placed next to each other, the inductive flows combine, moving the physical matter of the conductors as close as physically possible. When the conductors are placed so that their flows are in opposite directions, the inductive flows physically oppose each other, forcing the conductors apart if we attempt to push them together.

This effect, of course, is the same effect that occurs with magnets, which are made up of atoms whose nuclei have an excess affinity propensity that is too strong to be satisfied by the cloud of orbiting electrons that can exist without knocking each other out of orbit. Even though the individual nuclei cannot maintain their own orbital clouds, the nuclei still have an excess affinity propensity. This excess affinity propensity is structural and is satisfied by electrons which orbit the structure as best the structure will allow.

In a bar magnet, the orbiting electrons pass in one end of the physical structure, pass through the structure, and exit the other end, only to orbit around externally, evidenced by the easily measured external lines of force, and re-enter the first end once again to complete the orbit. When the ends of the magnet from which the orbiting electrons are exiting are pushed next to each other, the orbiting electrons push against each other, preventing the physical material from staying together. When the end from which the orbiting electrons are exiting is placed next to the end into which the orbiting electrons are entering, the orbiting flows combine, combining the magnetic material, causing the two magnets to come together to produce the shortest distance physically possible for the flows to satisfy the greatest excess affinity propensities of the nuclei of the atoms making up the magnets.

The strength of the inductive flow is always proportional to the strength of the primary flow. Thus, if we have a flow equal to "one" moving through the conductor, then we have an excess affinity propensity equal to "one" which will attract a number of ambient electrons equal to "one" into orbits around the flow. If we double the flow, we double the presence at any one point in the flow, we double the excess affinity propensity at any one point in the flow, and we double the number of ambient electrons attracted into the inductive flow, doubling the strength of the inductive flow.

Because there is always presence at any point in a flow of electrons, the effect of the presence is the existence of a stationary excess affinity propensity at any point in the flow. It is this stationary excess affinity propensity that replaces the less stationary affinity propensities of the electrons in the clouds of electrons orbiting the nuclei of atoms and it this replacement by a more stationary affinity propensity of a less stationary affinity propensity that results in combustion and the production of light.

Move a burning match next to a nonburning match and the nonburning match will begin to burn as the stationary affinity propensities of the match's emission field, its flows, replace the less stationary affinity propensities of the electrons that are orbiting the nuclei of the atoms that make up the surface of the nonburning match.

Stationary affinity propensities replace less stationary affinity propensities because the at rest motion of the less stationary affinity propensities is less of a potential and more of a reality.

8. Field Replacement

The reason that a physical description of combustion has eluded empirical science is because it monkey sees heat and monkey says energy. Energy requires input. Thus, on Earth, combustion is the chemical combination of oxygen and fuel atoms while in space it is the nuclear combination of hydrogen atoms.

Empirical science strains itself looking for the source of the energy when the energy is a simple release of potential energy, the electron's at rest motion, stored as a result of the affinity propensities of the electrons having overcome that at rest motion.

Because combustion is the mere release of the property of at rest motion, all that needs to occur to produce combustion is for the affinity propensities restraining the at rest motion to be removed. Because affinity propensity cannot be removed, it is a property of the electron itself, it has to be replaced. An affinity propensity has to be substituted for the affinity propensity holding the at rest motion of the electron in check to release the energy stored in that at rest motion.

The substitution of one affinity propensity for another is field replacement.

We can use field replacement to describe an endless number of physical phenomena, including chemical interactions, the thermal effect of electrons in a conductor and the production of light itself. We can see exactly what is happening in field replacement if we produce a flow of electrons and run it by an atom with orbiting electrons. At any point in a flow of electrons there is a presence, the existence of an electron at any point in the flow, and when there is presence, there is the effect of an electron at any point in the flow. Because electrons all have an excess affinity propensity that can be used up in satisfying the excess affinity propensities of other electrons, there is, at any point in a flow of electrons, an excess affinity propensity that is available to satisfy other excess affinity propensities in the environment.

The excess affinity propensities in the flow are stationary. As long as the flow exists, an affinity propensity will exist at any point in the flow. If there are electrons available in the ambient field, then those electrons will be attracted to the excess affinity propensity of the flow and will orbit that excess affinity propensity. Just as clouds of electrons will orbit the nucleus of an atom until all of the excess affinity propensity of the nucleus is used up and can not attract an additional electron into orbit, so will the stationary affinity propensities of a flow of electrons attract electrons into orbits at right angles to the flow until all of the excess affinity propensities of the electrons in the flow are used up and cannot attract an additional electron into orbit around the flow.

The stationary affinity propensity of the nucleus is the result of the at rest motion of the electrons in the units that make up the nucleus having overcome the at rest motion of those same electrons. The stationary affinity propensity of the nucleus is second only in potential, in permanence, to the stationary affinity propensities of the basic units that make up the nucleus. The stationary affinity propensity of a flow of electrons is not as stationary as the stationary affinity propensity of the nucleus because it is only as stationary as the flow itself. The electrons that make up the flow were either emitted in a directed field in the process of combustion, or have been placed in a directed flow as a result of potential differences in the environment, either arising naturally, or as a result of the hand of man in fabricated electrical circuits.

The orbital electrons, in contrast, are not stationary at all, and thus do not have stationary affinity propensities. The orbital electrons have their excess affinity propensities exactly balanced by their at rest motion. The stationary affinity propensity of either the nucleus or the flow of electrons has been able to overcome the at rest motion of the orbital electrons as the orbital electrons are attracted by these excess affinity propensities to the nucleus or the flow, but the electrons will remain in orbit only so long as there is not a more stationary affinity propensity that can satisfy the excess affinity propensity of the nucleus or flow of electrons that attracts and holds the orbital electrons in orbit.

Stationary affinity propensities will always replace less stationary affinity propensities because an electron has both at rest motion and affinity propensity and these two opposing properties are at war in the electron. If an excess affinity propensity attracts an electron, the electron's at rest motion might well be overcome, but only so long as the affinity propensity to which it is attracted exists.

If a more stationary affinity propensity starts to satisfy the excess affinity propensity to which an electron is attracted, then there is no longer an affinity propensity that will cause the affinity propensity of the electron to overcome its at rest motion and the electron's at rest motion will take over and move the electron away from the area at its at rest speed of light.

In atoms that have formed into matter, all of the excess affinity propensities of the electrons orbiting the atoms that make up the matter are used up in orbiting the excess affinity propensities of the nuclei of the atoms that make up the matter. Thus, when a flow of electrons, either in the form of light or electricity, strikes the surface of matter and begins to penetrate, the only electrons available to orbit the flow are the electrons that make up the orbital flows of the atoms. These electrons, conveniently, are not stationary affinity propensities, but rather are in the area only because of the excess affinity propensities of the nuclei. When the flows of electrons begin to penetrate the matter, they begin to satisfy the excess affinity propensities of the nuclei of the atoms that make up the matter, replacing the need for the orbiting electrons.

The orbiting electrons depart their orbits.

At the same time that the stationary affinity propensities of the flows are satisfying the excess affinity propensities of the nuclei of the atoms that make up the matter the flows are penetrating, the excess affinity propensities of the nuclei are satisfying the stationary affinity propensities of the flows. Thus, only a portion of the departing orbital electrons can be attracted to orbit the primary flows as inductive flows of electrons. Those that are no longer needed as orbital electrons and are not necessary to form inductive flows depart the physical matter.

We measure their departure as heat.

This explains why the misinterpretation of the photoelectric effect has produced the current Frankensteinian physics. When we discovered that light striking matter released electrons, we measured two types of electrons, energetic electrons released by blue light and less directed electron released by red light. We would expect that the brighter the light, the more its luminosity, the more energetic the electrons it released would be.

However, we already had the spectrum ordered in reverse, with blue light, the light that released high energy electrons, as high frequency light. Because low frequency red light released only low energy electrons, we could make a mindless mathematical relationship between the energy of the electrons released and the frequency of the light releasing them. This produced the concept of the wave/particle, the energy quantum called the photon, packets of light released by frequency or wavelength.

Knowing that blue is not high frequency light but low

frequency light allows us to see what is happening when blue light strikes the surface of the matter. The lower the frequency of the light, the easier it penetrates the surface of matter. Analogize light to radio waves as a reference point. Flows of low frequency blue light enter the surface, move between the nuclei of the atoms making up the surface and replace the affinity propensities of the nuclei holding the orbital electrons in place. These orbital electrons, already moving close to their at rest speed, are emitted at the surface of the matter. This is why they are measured as high energy electrons. They are actually emitted from the surface of the matter.

The higher the frequency of the light, as it moves from blue to red, the less it penetrates the surface, and the less internal field replacement occurs. However, the frequencies are breaking down at the surface, and the resulting heat is made up of ambient electrons, electrons that are moving randomly and which are thus measured as low energy electrons.

Of course, light is emitted in packets, but not as a result of the quantum shift in orbital electrons, rather as a result of the orbital electrons themselves being emitted.

An example of how flows of light attract excess affinity propensities is found in photographic film. Photographic film is composed of a gelatin which contains crystals that have an excess affinity propensity, silver halides that are measured to contain a charge. As flows of light pass through the gelatin, they are moving from the atmosphere, which has an abundance of ambient electrons that can form inductive flows around the flows of light, into the gelatin. In the gelatin, the flows of light seek to establish inductive floes. The areas around the flows have an excess affinity propensity.

The silver halide crystals have an excess affinity propensity as represented by their charge labeled positive in the polarity charged world of chemistry although no charge is necessary to the process. In addition, the crystals are capable of migrating in the gelatin, which is precisely what they do as they move toward the excess affinity propensities created by the flows of light.

When the flows of light stop, the silver halide crystals have been concentrated at areas in the gelatin where there were flows of light and because the excess affinity propensities are proportional to the flows, the crystals of silver halide concentrate in proportion to the strength of the flows that passed through the gelatin. As these crystals retain their charge after the flows of light cease, when developer is washed over the gelatin, the silver halide crystals combine into atoms of silver in proportion to their presence. If the film is removed from the developer after the concentrations have been coated with additional crystals but before the rest of the silver halide crystals that remain unconcentrated in the film are combined, a picture of the light appears that is shaded in proportion to the strength of the flows of light that passed through the gelatin.

An example of how excess affinity propensities replace other excess affinity propensities is found in chemical interactions. Chemical interactions are the result of manipulating the excess affinity propensities of atoms and molecules of atoms with the use of directed fields. If we take two compounds that are in balance with the environment, which is to say that their internal affinity propensities balance the at rest motion of the electrons that make them up, and the resulting structures are in balance with the field in which the compounds exist, say the surface of the Earth at seventy degrees, we can mix up a little brew. All we need is a field that is greater than the field in which the compounds exist.

We create such a field by starting a fire. If we have the compounds in a solution, heating the solution will affect the balance of affinity propensities of the compounds with their environment because by heating the environment in which the compounds are situated, we are increasing the stationary affinity propensities of the environment.

The atoms in the compounds all have clouds of orbiting electrons whose at rest motion is being overcome by the excess affinity propensities of their nuclei. When we heat the solution, we are increasing the affinity propensities of the environment in which the atoms are located and we are doing so with stationary affinity propensities.

Stationary affinity propensities replace less stationary affinity propensities.

To the extent that we increase the field in which the compounds exist, we are replacing the affinity propensities holding the electrons in orbit around the nuclei of the atoms that make up the compounds. The number of electrons released by field replacement is proportional to the field doing the replacing. As we measure these electrons as heat, heat in is producing heat out.

Thus, as the solution is heated, the atoms that make up the compounds lose their orbiting electrons in proportion to the strength of the ensuing field. If one of the compounds is made up of atom A and atom B, then atom A and atom B are held together by the attraction of the excess affinity propensities of their nuclei one with the other as mediated by the interference of the clouds of orbiting electrons that surround their nuclei. As the temperature of the solution increases, the orbiting electrons rejoin the ambient field as part of the increasing heat, the excess affinity propensities of the nuclei are replaced so that they drift apart, and the compound breaks down into its constituent atoms.

We then pop the second compound into the stew. If this compound is made up of atom C and atom D, the same thing happens that happened with the first compound, it breaks down into atom C and atom D. We now have four atoms in the solution, all with orbiting clouds diminished as a result of their replacement by the stationary affinity propensities of the heated solution, all existing separate one from the other because the field of the solution is also replacing the excess affinity propensities of the nuclei to the extent of the field.

We remove the field.

We take the pot off the stove and let it cool.

As the field diminishes, the nuclei, their excess affinity propensities no longer satisfied by the field that results from heating the solution, start to attract other excess affinity propensities in the environment. The nuclei begin to attract one another and also begin to attract electrons out of the ambient field back into their orbiting clouds.

When we started out, we had two compounds of two atoms each, each of which had alone found a balance in the existing field of seventy degrees. Now we have four atoms which are going to have to find a balance with each other in the field as it approaches seventy degrees. Each has a different potential difference with respect to the existing field, and thus different sized orbital clouds. Each also has a different level of excess affinity propensity in its nucleus.

The atoms attempt to come together as a result of the excess affinity propensities of their nuclei as moderated by the physical ability of the orbital clouds to knock orbital electrons out of each other's clouds of orbiting electrons. The nuclei reposition themselves into a structure that is in balance with the field of seventy degrees.

Where we had two compounds that had specific physical characteristics, we now have a single compound with completely different characteristics or perhaps one compound and a precipitate made out of one of the atoms that couldn't participate in the recombining process.

This is what is happening when we use trial and error to mix and match atoms and molecules of atoms to come up with the new compounds that we find to be useful in extending our range of survivability in physical reality. We manipulate the field in which atoms exist, take away some atoms and add some others, then return the field to its normal state to come up with something different.

We manipulate the field by increasing it, and in doing so we increase the stationary affinity propensities in which the atoms exist so that the increased affinity propensities of the field replace the affinity propensities holding the orbiting clouds of electrons in place, and as a result we are able to change the relationship of the atoms and molecules of atoms with each other.

A conductor is an arrangement of atoms that permits a flow of electrons to pass through it without altering the physical relationship of the nuclei of the atoms with each other. The nuclei of a conductor are themselves held together by their affinity propensities with a minimum of orbital electrons. One of the conductor's physical features, then, is that the nuclei can replace each other's excess affinity propensity and still maintain a distance one from the other without the aid of orbital clouds.

The structural difference between a conductor and magnetic material is that the nuclei of the conductor are not left with excess affinity propensities that would require additional orbital electrons. With the orbital clouds of the conductor's atoms at a minimum, the conductor has atoms with a maximum of space between the nuclei free of orbiting electrons. The electrons that make up the electricity is therefore free to move through the conductor at their at rest speed with a minimum of interference.

In generated electricity, as opposed to electricity produced in a storage battery, electrons are tipped into the conductor by passing the conductor through the flows of orbiting electrons around a magnet or though inductive flows created by a flow of electrons. While the electrons can travel at their at rest speed of light through the conductor, as diminished by the resistance of the conductor, which is a product of the affinity propensities of the nuclei and the physical existence of the orbital clouds that do exist, they will not be tipped into the conductor unless they have a place to go. A load on the conductor, a motor or a light bulb, creating a use for some of the electrons being tipped into the conductor, will produce a deficit of electrons, an excess affinity propensity toward which the at rest motion of the electrons can carry them.

This is what sets up the electrical flow in the conductor.

Flows of electrons moving through the conductor create an excess affinity propensity at each point in the flow as they pass through the conductor and this excess affinity propensity attracts electrons out of the ambient field to orbit the flow at right angles unless the conductor is insulated, in which case the electrons in the orbiting clouds of the atoms making up the insulation adjust their positions to accommodate the excess affinity propensities. If we increase the flow of electrons in the conductor, however, the excess affinity propensity of the flow will begin to replace the affinity propensities holding the nuclei of the atoms of the conductor together and the nuclei will move further and further apart, making the conductor more pliable than it already is.

If we place a load on the circuit that is greater than the conductor can carry, and increase the flow accordingly, then the flows will begin to affect the nuclei of the atoms themselves. With more electrons entering the conductor than can pass through the conductor, the electrons in the flows will begin to exit the surface of the conductor as heat.

Looking at this process under a mental microscope, we can see that as the flow of electrons replaces the excess affinity propensities holding the orbiting electrons in place, the orbiting electrons become ambient within the conductor. Some can be utilized by the flow of electrons through the conductor to establish inductive flows, but those inductive flows will then also begin to replace the excess affinity propensities of the nuclei, increasing the number of ambient electrons within the conductor.

Blocked by the crystalline surface of the conductor, and with the at rest motion of ambient electrons dominant over the electron's affinity propensity, these electrons begin to move in smaller and smaller spaces as their numbers increase. Pressure is brought to bear on the boundary surface of the conductor. The increasing mass of electrons, while in constant motion, is in constant motion in less and less of an area and no part of their affinity propensity is restricting the motion. As a result, the affinity propensity at the boundary begins to increase, causing the nuclei at the boundary to move apart.

A point is reached in the process at which the nuclei move far enough apart to allow a burst of electrons to be emitted. As soon as the burst is emitted, the nuclei close back up, but the pressures, if the flow through the conductor is sufficient, immediately build back up and another burst is emitted.

As soon as the bursts start, the rhythm of emission begins. The surface expands under the force of the build-up and emits electrons in the process. The surface then contracts shutting off the flow of electrons. The process starts to repeat itself with the frequency of repetition depending on the pressure for release.

As the atoms that make up the conductor begin to lose their cohesiveness and the conductor begins to break down, the electrons in the flow begin to take part in the emission.

The picture of light as waves rather than frequencies has produced a number of problems in visualizing what light is and how it moves. Because water waves are disturbances in a medium, they have a three dimensional quality on a two dimensional surface. This results in looking at waves not only by the frequency with which identical points, say troughs or crests, pass a certain point, but also by the distance between the troughs and the crests.

When Maxwell was attempting to visualize radio waves, produced by manipulating magnetic or electric flows of electrons into oscillating emissions, he was limited in his mental formulations to the picture of a three dimensional disturbance on a two dimensional surface. At any point on the surface of an expanding water wave, a wave point is passing, creating a mathematical frequency, and the point is bobbing up and down creating a mathematical amplitude of motion. Relating the frequency to the modulation, Maxwell decided that a radio wave, and indeed a light wave, was merely the interaction of electric and magnetic forces acting at right angles to each other. The source of the disturbance, an electromagnetic oscillation, was propagated through the aether by the electromagnetic interaction of these two forces at right angles to the direction of travel.

At the time, everyone was so astounded that something could happen at a distance, that information could come out of thin air, that an entire industry grew up around mediums who claimed to obtain spiritual messages from the afterlife. How this might occur was eminently more comprehensible than Maxwell's idea that light moved forward in a medium as a result of the interaction of electric and magnetic fields, but nothing was more rational than Maxwell's unification in a single visualization of light and radio waves as different frequencies of the same thing.

And that's all they are.

There is no wave and there is no amplitude. Amplitude modulated radio waves are simply a different manipulation of lower frequencies than the higher frequency modulated radio waves. With two hundred years of wave thought behind us, two hundred years in which the terminology was devised to describe the technical innovations that fill our lives, technical innovations which were monitored by detectors that were designed to visually display waves on screens, it is hopeless to expect a change in terminology.

However, if we want a change in understanding so that we can actually comprehend what is going on with light, if we actually want to know the structure of light, then we are simply going to have to give up the wave. This is not even easy to accomplish mentally. Along with the wave picture, we have inherited the up slope, down slope picture. As anyone familiar with the beach knows, you bob up on the incoming side of the wave and bob down on the back side of the wave. Waves have an up and a down, or in Maxwell's visualization a right and a left, up and down and all the points in between. Whatever the picture, it is a picture of two separate processes that produce two separate effects. The oscillation that occurs on the face of the sun as it undergoes combustion, or on the surface of the copper conductor as it begins to glow red does not produce an up and a down or a left or a right.

The boundary atoms at the surface of the conductor begin to move apart and then move together again in a rhythm reminiscent of Maxwell's concept of the interaction of electric and magnetic forces. Here the interaction is between the affinity propensities holding the nuclei of the boundary atoms together and the affinity propensities building up in the electrons whose at rest motion is attempting to break through the physical surface created by the boundary nuclei.

The electrons are emitted in packets with each oscillation, the packets themselves held together by the affinity propensities of the electrons that make them up. Before the surface of the conductor can begin to emit these packets, an enormous number of electrons have to be built up, but then when we are dealing with particles that are only large enough to define nonexistence, an enormous number can fit in a ridiculously small space.

With a bare copper conductor, once the emissions begin, the presence of the flows causes field replacement to occur in the atoms surrounding the conductor. If these atoms contain oxygen, as they would if the conductor were sitting on a laboratory bench, then the electrons that made up the orbiting clouds of the oxygen atoms would depart at their at rest speed of light, causing further field replacement at the surface of the copper. This field replacement would result in the boundary atoms beginning to come apart, increasing the frequency of emissions, perhaps causing the copper wire to erupt in a burst of light as it disintegrated.

The same result could be accomplished by increasing the flow way above the wires ability to contain it. In either case, field replacement results in the emission of frequencies of light. These frequencies depend solely on the rate at which a substance is undergoing field replacement, which is combustion.

Once we have ignition, the point at which light is being emitted in the process of combustion, field replacement will continue as long as there are orbital clouds that the field can replace. This process is most easily seen in the match. A match has a head which is made up of a substance that can be ignited by friction. Ignition is the beginning of combustion and is itself a process that needs description.

In the copper conductor just incinerated, the primary source of electrons that are emitted in the process of field replacement is the electric flow. The electric flow drives the emissions which produce the heat and light.

Add some resistance to the conductor and put it in a vacuum tube and we have a light bulb.

When field replacement occurs in an area made up of contiguous atoms which contain abundant orbital flows of electrons, then the field replacement can begin to occur in all directions. The failed concept of an atomic chain reaction provides a picture of what actually happens when contiguous atoms begin to emit. In the classic chain reaction, neutrons cause fission which releases more neutrons which causes more fission which releases progressively more neutrons in a cascading effect that eventually destroys the whole world. In ignition, matter goes from a state of balance with respect to the environment in which it exists to a state in which it is capable of consuming itself in the process of combustion.

When the match surface is scraped across an abrasive

surface, the boundary nuclei are being physically torn apart one from the other, immediately exposing the excess affinity propensities of the nuclei that held them in place. These excess affinity propensities immediately affect the clouds of orbiting electrons of the contiguous atoms, and the orbital clouds begin to depart from contiguous points in expanding spheres of emissions.

Each atom becomes a miniature emitter immediately. All of these expanding spheres contain excess affinity propensities, and as they expand, they intersect, increasing the excess affinity propensities of the points around them.

These points all contain atoms with orbiting clouds of electrons, normally atoms, or molecules of atoms that have the largest clouds possible to make the matter ignite easily, and these atoms or molecules of atoms immediately become emitters. Their expanding emission fields join the existing expanding emission fields causing other atoms to emit, and the process continues to increase until the entire pile of contiguous atoms ignites and begins to emit as a unit.

Unlike the copper wire, where there was a surface to oscillate, the electrons being emitted at the surface of a match head do not have a boundary condition to overcome. They have an area condition to overcome.

When the match head ignites, there are simply too many ambient electrons in the area for all of them to leave in an orderly fashion. The combustible surface itself becomes the boundary. The electrons have to organize themselves on the surface created by their collective affinity propensities before they can escape.

The result is the same as with the surface of the conductor, with the surface expanding from the internal pressure, packets of electrons being emitted which allows the surface to contract, and the surface then re-expanding to emit another packet.

While this process is going on, the excess affinity propensities of the ambient electrons crowded together in the area of ignition awaiting an opportunity to leave at their at rest speed in the structured emission of light begin to replace the affinity propensities of the atoms of the less volatile wood or paper stem, releasing the electrons from their orbital clouds. The atoms or molecules of atoms that make up this matter also begin to undergo the same process that occurred as a result of friction, with the addition of the oxygen molecules at the boundary providing additional points of expanding emission fields to keep the process going.

The match continues to burn so long as there are boundary atoms to provide electrons from their orbital clouds to supplement the electrons from the orbital clouds of the atoms that make up the matter in the match stem undergoing field replacement, combustion.

The effect of field replacement as combustion can be seen when we take an ignited match and hold it up against another unignited match. We are all aware of the fact that matches burn when they are close to us, and don't burn when they are at a distance.

A burn is simply the process of field replacement that occurs in the material that makes up the skin when a source of strong emissions is brought too close to the skin. The frequency being emitted by a burning match, when brought next to the skin, is sufficient to replace the affinity propensities holding the boundary atoms at the surface of the skin together. As the affinity propensities of the nuclei are replaced, they move apart. As they move apart, the flows of electrons from the match penetrate the surface of the skin and start to replace the affinity propensities of the nuclei holding orbiting clouds of atoms together beneath the surface of the skin and these orbiting clouds in turn depart, causing additional field replacement. If you hold the match on the surface of the skin long enough, then field replacement will continue until the atoms have broken down, completely destroying the surface.

The flame does not contain a conglomeration of high, medium and low frequencies and all the frequencies in between. Nor is it a conglomeration of all the waves that can exist on the surface of a turbulent ocean.

The flow contains a single frequency determined by the rate of oscillation as its surface emits electrons.

This single frequency produces the field replacement that we measure as the effect of the light, whether we see it with the receptors in our eyes or feel it with the sensors in our skin, or both.

This seems so obvious that it needs no explication, and

yet we visualize the light component of the emissions that result from combustion as luminosity, we see that luminosity as made up of all colors in the spectrum, we see the luminosity portion as one portion of an overall electromagnetic spectrum which also has a different component that causes vibrations in matter that result in the matter acquiring heat, and these vibrations occur quite independent from the light.

We are dealing with something so deceptively simple that we literally can't see it!

When matter undergoes combustion, the at rest motion of the particles that make it up is overcoming the affinity propensities holding the same particles together. The particles are being released at their at rest speed of light simply because there is nothing left to hold the electrons in orbit. The excess affinity propensities that were holding the electrons in orbit have been satisfied by the more stationary affinity propensities in the flows of electrons.

Field replacement allows the at rest motion of the orbital electrons to become dominant. The at rest motion causes the electrons to depart in packets whose size is determined by the rate of emission which in turn is determined by the number of ambient electrons attempting to leave the area. A burning match might have a minor number of electrons attempting to leave the area and thus putting a minor amount of pressure on the boundary area of combustion while the surface of the sun, where the basic units themselves are breaking down into their constituent electrons, will have incredible pressures being built up that cause the packets to be released at a very rapid rate.

Because combustion is field replacement, and field replacement is heat, the amount of heat is determined by the number of electrons crowded together into a single packet and the number of packets that pass a given point at a given time. The former is intensity, the latter frequency. When the frequency at which the packets are released reaches a high enough level, we "see" the heat with the organ that evolved to perceive a useful range of frequencies, the eye, as light but only when the intensity has diminished to the point that the surface of the eye won't be destroyed by the intensity.

When we take a lighted match and hold it some distance from an unlighted match, the effect of the lighted match on the unlighted match might be no more than to provide enough light to see the unlighted match. The emissions from the surface of the burning match are moving out in every direction over the surface of an expanding sphere, and as they are doing so, they are diminishing inversely with the square of their distance from the burning match.

When the second match is held four or five feet away from the burning match, then the electrons that make up the expanding sphere of emissions from the first match merely bounce off of the surface atoms of the second match. There is not a sufficient amount of presence in the emissions to affect the physical structure of the second match, and no affinity propensities are replaced in the second match. If the second match were held a mile away from the first match, then there would be no emissions left to bounce off the second match, the emission field having expanded to the point that it broke down with the electrons that made it up becoming a part of the ambient field.

Light dissipates out of existence.

Between the lighted match and the unlighted match exists an emission field which diminishes inversely with the square of the distance. This emission field has affinity propensities at any point in its existence, with the strongest affinity propensities being at the surface of the combustion process of the burning match, and the least affinity propensities at the surface of the noncombusting match.

We can increase the affinity propensities affecting the noncombusting match by moving it deeper into the expanding emission field of the combusting match (or if we were using a filament of some sort for the first match, we could accomplish the same thing by increasing the emission field of the filament, an important aspect of how matter interacts with other matter in space where it moves about in fields that increase and decrease as a result of the movement). As we move the noncombusting match and the combusting match closer together, we are increasing the field in which the noncombusting match exists, and as we do so we are increasing the affinity propensities in which it exists.

The emission field of the match is composed of flows of electrons which have presence so that at any point in the flow there is an affinity propensity that affects the affinity propensities of the nuclei at the boundary of the match. These affinity propensities are increasing in direct proportion to the square of the distance as the distance is reduced, so that before long, the affinity propensities of the burning match are beginning to replace the affinity propensities holding the nuclei of the surface of the nonburning match together and are beginning to penetrate that boundary causing field replacement in the clouds of orbiting electrons of the atoms making it up.

These electrons in turn become points at which expanding fields occur and when the match gets close enough for the expanding fields to release enough electrons to create an overwhelming presence at the surface of the match, the match ignites, bursts into flame, and begins to combust on an orderly basis, emitting its own expanding sphere of light in the process.

Combustion is merely the release of the electron's at rest motion as a result of the replacement of the affinity propensity holding that at rest motion immobile, and energy, light, is the result of immobilized electrons regaining their at rest speed, the speed of light.

9. The Structure Of Light

We now have a clearer picture of light than we could create from a bunch of facts held together by unexamined assumptions used to form uncheckable concepts.

We know that light is something, that it moves from one place to another. We can discard the misconception that light is a wave phenomena for several reasons. Young's two-slit experiment merely tells us that light is recombining in some way, not moving like water waves. In addition, light as a wave phenomena requires that there be a medium to disturb and there is no measurable medium to disturb, especially between the sun and the Earth.

Further, as opposed to our process of constructing detectors to see the waves we believe are there, everything that we do on a technical basis, whether it be detecting starlight electronically, operating our television sets at a distance or opening our doors automatically, uses light as a particle and not a wave. We can therefore operate with the concept that light is what Newton originally thought it to be, a particle.

We know that light has frequency, not necessarily from the knowledge of light itself, but from the knowledge derived from producing lower frequencies than light, radio waves, microwaves and the like which are not waves but oscillations produced by manipulating flows of electrons.

These frequencies are produced in a manner that allows us to measure the frequencies, and while we cannot directly measure the frequency of light, we know that the combustion process is an oscillatory process. Because we know that light comes in colors, we know that light has different frequencies. Because light has different frequencies, and because different frequencies appear in nature as light bounces off of different matter in reality, we know that the interaction of light with matter changes the frequency of the light.

We know that light has an inherent speed because we have to restrain it to slow it down and once the restrained light is released, it speeds right back up. Combining this knowledge with the knowledge that light is a particle tells us that light travels in flows made up of particles.

Putting frequencies in flows of particles that can interact with matter produces a clear picture of the frequency changes that occur from interacting with the surface nature of the matter, pigment, from interacting with the nature of the matter, a prism, or even from interacting with the motion of matter, the effect of bouncing a frequency off of a moving object to detect its direction and speed.

We know that light expands over the surface of an expanding sphere because it is emitted in every direction, and because we know the relationship of the expanding sphere with distance, we know that light diminishes uniformly over this surface as it expands inversely with the square of the distance over which it travels.

These facts, combined with the knowledge that light travels as flows of particles, tell us that the flows of light that are emitted from the surface of combustion are dividing as they expand. Each packet of light that is emitted breaks up into smaller packets as the light moves away from the source of emissions and as each packet breaks up into smaller packets, the smaller packets arrange themselves uniformly over the surface available for their expansion.

As expansion affects the intensity of the light, but not the frequency, we know that something is holding the light that makes up each frequency into a cohesive structure. It takes the interaction of light with matter to alter its frequency.

And finally, because we know that light diminishes uniformly as it expands, we know that there has to be a mechanism that regulates the expansion of light. If there weren't a mechanism that regulates the expansion of light, then it would not diminish uniformly as it expanded.

Once we realize that the only way that light could expand uniformly is if it were kept from overexpanding, we are immediately reminded of Young's two-slit experiment in which light is recombined out of areas into which it has been allowed to overexpand. The realization that there is something recombining light in Young's two-slit experiment suggests that whatever could recombine flows of light could also regulate its expansion by limiting its overexpansion, allowing light to diminish uniformly as it expanded the way we measure it to do.

A close look at Young's two-slit experiment, then, should provide us with a description of what is recombining the expanding flows of light and how it is recombining them, thus giving us a clue as to the overall structure of light.

Young's two-slit experiment is simple in execution and comprehension. If we have a single source of light, and we allow that light to pass through two slits in a barrier, when we place a screen on the other side of the slits to collect the resulting light we find bars of light and no light on the collection screen.

For two hundred years we have mindlessly said that the areas of no light are similar to the areas in intersecting water waves where the troughs of one wave cancel out the crests of another wave. We have blindly repeated the statement that the areas of darkness are interference patterns, the troughs of one light wave interfering with the crests of another light wave to cancel each other out.

Once we made this leap of faith, we could go crazy with our mindless mathematical mayhem. If we know the distance the slits are apart, and the distance the collecting screen is from the slits, we can compute the wavelength of the light. If we wanted to get more exact in our computations, then we could etch lines in glass so that the distance between the etches would allow the determination of wavelength by the colors that passed through the etches.

We are merely measuring our assumptions, so it

doesn't really matter what we come up with as long as we can agree on the outcome and avoid arguing the assumption. This sort of monkey see, monkey say business is the bread and butter of empirical science where assumptions are only as good as the capacity for mindless measurement they generate.

The misconception that the dark bands on the collection screen were actually the existence of light, the cancellation of troughs and crests that were measurable in water waves, and that light was therefore just like water waves gave us so much industry computing the resulting consequences of the misconception that we didn't have to bother thinking ever again about the absurdity of the belief that nothing, darkness, was actually something, light, in fact, twice as much something.

We didn't have to bother with the obvious physical realities presented, that when the peaks and valleys of water waves cancel each other out, there is still water, and when the light falls on the collection screen, there is no light between the bands of light.

This fact in physical reality, that the absence of light means there is no light, rather than twice as much light, is difficult for empirical science to grasp, but it seems to be quite self-evident.

No light does not measure out as light.

No light means that there is no light to measure!

If there is no light, then something has happened to

remove the light from where there is no light because if the slits are adjusted in the experiment so that they are no longer the same distance from the source of light, then the light magically appears where before there was no light. Well, it doesn't magically appear. What happens is that the light that was concentrated in the bars of light is redistributed over the entire surface of the collection screen.

To empiricists, the disappearance of the interference patterns showed that light waves were arriving at the collection screen out of phase, the result of the light traveling different distances at the same speed, with the troughs and crests hitting simultaneously and thus producing uniform light.

This result had even more fantastic results than the interpretation of the absence of light as the presence of light and the subsequent computation of wavelengths which were actually computed backwards, with red and blue mixed up and placed at the wrong ends of the spectra.

At the end of the nineteenth century, the luminescent aether, the movement of light through space, had been replaced by the just plain aether, the medium through which light waves were propagated. As light was propagated through a medium, a beam of light sent through the aether at right angles to the direction of the Earth's movement in that aether would travel at a different rate, and thus would take longer to travel the same distance than light sent in the direction the Earth was traveling would take. Traveling at different rates over identical distances would have the same result as the light in Young's two-slit experiment which travels different distances at the same rate had. The light would, if an experiment could be devised to collect it, arrive out of phase and thus incapable of interfering.

Michelson and Morley reasoned that the movement of the Earth in space, its absolute speed and direction, could be computed by using the extent to which the light waves that arrived at a collection screen were out of phase as evidenced by the lack of the dark bands, the interference patterns. The result of this experiment, upon which the entire edifice of twentieth century empirical science rests, was nothing. The interference patterns appeared on the detector as if the experiment had been performed along the classical two-slit lines with the light traveling the same distance at the same rate.

An examination of how light moves demonstrates why the light behaves exactly as it does behave in Young's two-slit experiment. When the light is emitted, it begins to expand over the surface of an expanding sphere. If nothing impedes its expansion, it will strike the collection screen diminished inversely by the square of the distance of the screen from the source of light. When a barrier containing a slit is placed between the source and the collection screen, what happens? The light passes through the slit in the barrier and begins to expand over the surface of an expanding sphere, an expanding sphere with a new center of expansion, and one made up only of the light that passes through the barrier.

When the light passes through a single slit in the barrier, it begins to expand over the surface of an expanding sphere until it falls uniformly on the collection screen. When we cut another slit in the barrier, we have twice as much light expanding over the area between the screens, but the light is intermingling. Instead of the combined light falling on the screen with twice the intensity of the light from a single slit, it falls on the screen in bands of no light separated by bands of light that are twice the intensity of the light that would fall on the screen if the light from both slits could fall uniformly on the screen. We have taken two identical areas of light from an expanding sphere of light and given them new centers of expansion. If we want to increase the effect, we can overexpand the light by passing it through a slit or a pinhole before we pass it through the two slits to be further overexpanded.

Light from an identical source is passed through two identical slits identical distances from the source of light and allowed to intermingle. The flows of electrons from both slits are identical. The packets of electrons that are its individual frequency were emitted by the same atoms undergoing field replacement at the same rate so that while each packet contains untold gadzillions of electrons, the number in each packet emitted from the surface at each instance is the same. The surface is oscillating at the same frequency so that the frequency of both sets of light flows is the same. In addition, both light flows were originally emitted over the surface of the same expanding sphere. The light was intercepted at two points on the surface of that expanding sphere by identical slits in the barrier at identical points in that expanding sphere and are therefore two expanding spheres that are re-expanding so that at any point at which the light from one slit is intermingling with light from the other slit, the light is identical.

The intermingling light is made up of individual flows of electrons and a flow of electrons has presence. At every point in a flow of electrons, there is always an electron and that point therefore has the property of an electron even though it is not being occupied by the same electron. One of the properties of an electron is affinity propensity and therefore at any point in a flow of the electrons that make up the expanding flows of light in the two expanding spheres there is an excess affinity propensity that is proportional to the strength of the flow, the intensity of the light.

If we place two conductors carrying a flow of electrons in the same direction side by side, the inductive flows will move the conductors together. The inductive flows combine the conductors because the separated inductive flows interfere with one another, knocking electrons out of the inductive flows. As electron are knocked of the inductive flows, excess affinity propensities appear in the flows. Because the flows are contiguous, these excess affinity propensities attract each other, moving the flows together and carrying the physical matter of the conductor along with the flows. As the conductors close the gap, more interference creates a greater loss of electrons in the inductive flows accelerating the effect of the excess affinity propensities of the flows in closing the gap. The conductors snap together much as magnets snap together in a process that also combines flows of electrons.

Flows of electricity contain massive numbers of electrons. We are creating the flows and we can control their magnitude by manipulating the size of the generators producing the flows together with the loads producing the excess affinity propensities that attract the electrons in the flows. We can detect these massive flows with ease. They can physically grab us, turn our muscles into vibrating marble, and in some cases, strip the skin from our bodies. Small flows deflect magnets or other flows, making detectors extremely simple.

When we look at light we don't see electricity. We don't see anything like electricity. Electricity moves in a conductor. Light doesn't need a conductor. Electricity requires a circuit. Light doesn't have a circuit. Electricity has a charge. Light doesn't have a charge.

When we monkey see electricity, we can't monkey say light!

However, once we realize that movement is a result of the property of at rest motion of the electron, we realize that the electric conductor and circuit are no longer a bar to us seeing light and electricity as a flow of the same thing. Both flows of electricity and flows of light are directed flows of electrons moving at their at rest speed. When we create the flows of electricity, we have to provide a path and purpose for the flows. The flows produced by combustion, however, are directed by the combustion process itself and are structured as a result of that process. The electrons are released in packets over the surface of an expanding sphere and those packets make up flows whose frequency is determined by the rate of combustion. The rate of combustion also dictates the number of electrons that go into each packet, which is the intensity of the light flows produced at the surface of combustion, and as those packets expand over the surface of the expanding sphere, the number of electrons in each packet diminishes so that the intensity of the light diminishes inversely with the square of the distance over which it moves.

The flows, however, have presence, and the presence produces an excess of affinity propensity that attracts clouds of orbital electrons, electrons which because they are orbiting a flow, orbit the flow at right angles to the flow. This is what produces the inductive flows in electricity and this is what produces the inductive flows associated with light. These inductive flows are measurable on the surface of heated elements on Earth as magnetic or electric charges. The inductive flows of electricity are what produces electricity in generators and what are insulated against in electric wiring.

But the inductive flows in light are what balance the charges associated with the flows of light so that light appears to have no charge at all to our detectors until it strikes the surface of matter, and becomes the photoelectric effect either by low frequency blue light field replacing orbital electrons in the nuclei of the atoms that make up the matter or by higher frequency light breaking down the surface and itself into ambient electrons.

The question then becomes not a question of how the flows of light differ from the flows of electricity, but rather how the inductive flows generated by the presence of light interact with the flows of light when those flows are expanding over the surface of an expanding sphere, and breaking down into smaller and smaller flows in the process, because that is the structure of light, the interplay of expanding flows of light with the inductive flows produced by the expanding flows attempting to recombine the expanding flows.

Inductive flows are flows of electrons in circular orbits around a primary flow of electrons, be it in light or in electricity in a conductor. We can actually see the effect in the flows of electrons that make up the external orbits of magnetic material and which we measure as magnetic lines of force. These flows of electrons, moving from one end of the magnet in a arc that accommodates the at rest motion and the affinity propensity of the electrons involved, produce their own inductive flows which can be measured by having them pass vertically through a plane carrying iron filings. The iron filings will line up in circular paths tracing the inductive flows of the magnetic lines of force.

If we run a conductor carrying an electric flow in the same direction as the magnetic lines of force through an identical plane carrying an identical pile of iron filings, we will get an identical result. If we take a miniature compass needle, a magnet with one end marked north, the needle will be deflected by the magnetic lines of force in the same manner that it is deflected by the electric flow. The only difference is one of degree, with the electric flow producing a more defined inductive tracing than the magnetic lines of force because of its greater presence. We certainly wouldn't suspect the inductive flows around the magnetic lines of force unless we could measure them. The inductive flows, however, maintain the cohesiveness of the magnetic lines of force, and with respect to the conductors, actually cause the physical matter of the conductors to come together if the current flows in the conductors are moving in the same direction.

We can keep the inductive flows straight by using the right-hand rule. If we point the thumb of the right hand in one direction, the fingers will curl at right angles to the thumb. The thumb represents the direction of the current flow while the fingers curl in the direction of the circular inductive flows. When two conductors with two flows and two inductances are brought contiguous to each other, the at rest motion of the electrons that make up the inductive flows are going to orbit the excess affinity propensities of the primary flows. With two inductive flows attempting to orbit two primary flows moving in identical directions, the two primary flows will begin to move together.

If we wanted to physically measure inductive flows in light, we would have a problem. If the light were of an insufficient intensity to penetrate the plane carrying the iron filings, then the flows of light would simply bounce off the plane. No inductive flows could be established on the face of the plane. If the light were of a sufficient intensity to affect the position of the iron filings, it would burn the plane up or melt the iron filings, or both, thus defeating the experiment.

We don't have to perform experiments with iron filings to visually see the inductive flows at work in light, however, because the dark lines on the collection screen in Young's two-slit experiment and the concomitant brighter contiguous bands show that the inductive flows associated with light have recombined the light to the extent possible.

When we intermingle the two expanding spheres in Young's two-slit experiment, at any point at which the two expanding spheres intersect, there are flows of light from two individual overexpanding spheres overexpanding parallel with one another.

A flow of light is not parallel from the slit to the collection screen, but rather at every point in the expanding sphere. Every point on an expanding sphere is a point of expansion so that contiguous points produce flows which are expanding parallel to other contiguous points. The recombination that occurs when expanding spheres of light are intermingled is the actual recombination of congruent flows, because two expanding spheres of light are providing identical points of expansion at the same place in space.

The structure of light, then, is created by the interplay

of the flows of light overexpanding into the areas available for their expansion, an expanding sphere, being recombined to the extent possible by their associated inductive flows. At any point on the surface of an expanding sphere there is an expanding sphere and the flows of light from each of these expanding spheres intermingles with the light from contiguous expanding spheres. The inductive flows can recombine expanding flows of light from contiguous spheres so long as the frequency of the light is synchronous, the term that equates with phase in wave theory and means that the frequency are both the same length and those lengths are aligned, and of the same intensity.

The inductances recombine the expanding flows because there are more flows in the area than the amount of affinity propensity at that point in the expanding sphere will permit. Inductances form in proportion to the affinity propensities of the flows, but light expanding on the surface of an expanding sphere expands into smaller flows than the inductive flows produced by the total flows will allow to remain apart and the inductive flows immediately recombine the overexpanded flows. However, because the recombination occurs after the overexpansion, it never recombines the light to the point that it began to overexpand. As soon as the light has been recombined, it immediately overexpands once again.

As the process repeats itself a little further from the source of light each time, the process allows the light to both expand and to do so uniformly.

The interaction of the expanding flows of light with their associated inductive flows produce a structure that can retain its individuality so long as there are sufficient electrons in the ambient field to form the necessary inductive flows. This is why light from two or more sources can retain their identity in the environment. We see many different expanding spheres of light when we watch a television screen and we can pass two beams of light through one another and see their individual effect on collection screens. It also explains why one light can wash another out. If one source of light is strong enough, it will use up the electrons in the ambient field to form its own inductive flows leaving none for the weaker light, prevent that light from forming inductive flows and that light will never recombine, it will simply expand out of existence, its electrons perhaps becoming part of the inductive flows of the stronger light.

In Young's two-slit experiment, identical flows of light, light with the same frequency and intensity end up occupying the same space. As long as the center of the expanding spheres is the same distance, the light remains identical. Because there is twice the light at each point in the intersecting expanding spheres of light, when the expanding flows from both expanding spheres are recombined, the light from both expanding spheres is recombined in accordance with their presence. This presence is twice the presence of the individual flows of light, and the inductive flows therefore recombine the expanding flows back to the point that it would have been recombined had it not been divided.

This leaves areas of no light out of which light has been recombined.

There are other areas in physical reality in which we see inductances recombining identical frequencies. As peacock feathers shimmer and waver, reflected light expands off of different surfaces with identical frequencies, and, as the expanding flows momentarily recombine, the light glistens with a brightness that surpasses the brightness of the colors reflecting off of the individual surfaces. As the source and our eyes move with respect to one another, the iridescent colors come and go as we see recombined light flash in the midst of nonrecombined light. Another more problematic area in physical reality where flows of light from the same source recombine is glare. We have all been blinded when sunlight glints off the chrome on a car in front of us or off a patch of ice on the road to produce a light that is brighter than the light reflecting off the chrome or ice. The expanding spheres of light reflecting off of an irregular surface will intermingle and recombine into flows of light that are much brighter than the light that would normally be reflected. We even use the effect to produce popular technology, shaping the beam of flashlights with concave reflectors.

When Michelson and Morley conducted their experiment to determine the speed and direction in which the Earth was traveling through the aether, they split a beam of light from a single source and sent it down two paths. The first path went the diameter of the circular experimental platform, was reflected back with a mirror to the center and then reflected from the center to the detector on the right side of the platform. The second path went to the center of the platform, was deflected to the right side, then reflected the diameter of the platform to the right side to be recombined with the other beam at the detector, a device designed to measure the existence or nonexistence of an interference pattern.

The pair expected the interference patterns to disappear when the split beams were recombined and reached the detector because, although the split beams were traveling the same distance within the physical bounds of the platform, Michelson and Morley expected the movement of the Earth with respect to the aether to make the split beams travel at different rates because the platform itself was moving with the Earth.

The platform was circular so that it could be rotated in any direction, and, so the logic went, sooner of later one beam would be traveling in the direction that the Earth was traveling and the other would be traveling at right angles to that direction so that one beam would be traveling at a faster rate than the other.

No matter which way the platform was rotated, however, the detector got the same result, bands of light and no light. As a result, everyone concluded that because the light was obviously moving at different rates through the aether, an indisputable scientific fact, the physical distances of the platform had to be changing to compensate for the different rates at which the two beams of light were traveling.

If the light speeded up, then the physical dimensions of platform expanded accordingly. If the light slowed down, then the physical dimensions of the platform shrunk accordingly.

What other conclusion, an empiricist named Fitzgerald wanted to know, was one to draw?

The split beams, however, were produced by the same source and thus had identical frequencies. As they expanded over identical distances, when there were once again intermingled before reaching the detector, they had identical intensities. When the two requirements for recombination were met, frequency and intensity, the light recombined without reference to hypothetical media which would hypothetically alter hypothetical rates which then required hypothetical changes in time and distance to accommodate the continued existence of a simple physical phenomena.

In the Michelson and Morley experiment, the bars of light and no light, the interference patterns, are the result of the light reaching the collection device being synchronous in frequency and of the same intensity because the same light has been split and the expansion has occurred over identical distances. The interference patterns are the result of the experiment's apparatus and not the motion of that apparatus, and thus the result is, or at least it should have been, expected.

If Young's two-slit experiment had been performed perhaps four decades later, after the existence of inductance became known, I suspect that our technological development might have taken a somewhat different course. Viewing the result of the experiment as the recombination of light by its associated inductive flows reveals much more than the structure of light as the interplay between its expansion and the associated inductive flows regulating that expansion.

The inductances associated with the expansion of light have a far greater consequence than causing light to diminish uniformly as it expands over the surface of an expanding sphere.

If we look closely at what is happening when light expands over the surface of the expanding sphere, we can discern a series of relationships, factual congruities, actually, with enormous consequences.

First, we know that light diminishes inversely with the square of the distance from its source. This fact is geometrically established by the relationship of the area of a sphere to its radius, with the areas of spheres with the same center being related by the square of their distances to the center.

This inverse square measurement is the exact same measurement we have for matter falling toward the surface of the Earth and is thus the exact same measurement we have for gravity. The fact that matter accelerates with the square of the distance over which it falls means that the force that is making it fall is diminishing inversely with the square of its distance from the point toward which the object is falling.

At any point in an expanding sphere of light, the amount of light present at that point can be determined with reference to the square of the distance of the point to the source of the light. We cannot know how much light is present at any particular point, the light's overall intensity, because we don't know how much light was emitted at the source. However, the measurement allows us to know that the amount of light at any point in an expanding sphere of light is related to the amount of light at any other point in an expanding sphere of light by the square of the distances of each to the common source of the light.

The amount of light is the flow of light at a specific point. A flow of light at a specific point has a presence and that presence produces an excess affinity propensity. The excess affinity propensity is proportional to the flow of light so that at any point in a flow of light there is an excess affinity propensity that is inversely proportional to the square of its distance from the source of the flows. Because the excess affinity propensity at any point in a flow of light produces an inductive flow, at any point in a flow of light there is an inductive flow whose strength is inversely proportional to the square of its distance from the source of light.

Light expands over the surface of an expanding sphere. At each point in an expanding sphere of light, the packets are splitting up into smaller packets, which is to say the intensity of the light is becoming less. The frequencies are identical, they just contain progressively less electrons held together by their affinity propensities. The light from each expanding point is intermingling with the light from the contiguous expanding points. Because the strength of the associated inductive flows that are recombining the expanding flows of light are inversely proportional to the square of their distances from the source of light, the force with which these inductive flows recombine the light is also inversely proportional to the square of their distances from the source.

This recombination of light by its inductive flows is a mechanism that occurs at any point in an expanding flow of emissions, and because it occurs at every point on the expanding sphere of light, it is the mechanism that regulates the uniform expansion of light over the surface of the expanding sphere.

The force that the mechanism produces is always inversely proportional to the square of the distance from the source which is the exact same measurement we have for falling objects.

Basic units in the nuclei of atoms of matter caught up in an expanding sphere of light are affected by this mechanism. As the mechanism recombines the light, the force of recombination is back in the direction from which the light is expanding. Acting on the individual units that make up the nuclei of atoms, the mechanism causes these basic units to move back toward the source of emissions. As each mechanism back toward the source of emissions becomes progressively stronger by a factor equal to the square of the distance from the source of emissions, the basic unit in the nuclei of an atom moves at an increasing speed as it moves to each stronger mechanism, and in fact accelerates with the square of the distance over which it moves toward the source of the emissions.

So long as a falling object is unimpeded, the basic unit accelerates with the square of the distance over which it moves. It makes no difference whether the basic unit is tied together with fifty or a hundred other basic units, all atoms accelerate at the same rate because the mechanism acts on each individual unit to move that basic unit.

It is only when an attempt is made to stop the movement of atoms with different numbers of basic units held together in their nuclei that the difference in the number of units tied together in the nucleus of an atom has an effect.

Because all units tied together in a nucleus of an atom have to be moved at the same time, moving an atom with a hundred basic units in the nucleus against the force of the mechanism means that there are a hundred recombining mechanisms that have to be overcome at the same time, twice the effort it takes to move an atom with fifty basic units in the nucleus against the force because in that case only fifty recombining mechanisms have to be overcome at the same time.

Because we measure the force it takes to work against these recombining mechanisms as weight, different atoms have different weights when they are held against the effect of the recombining mechanisms while all atoms regardless of weight accelerate at the same rate when they are allowed to move freely within the recombining mechanisms.

The effect of the mechanism that regulates the uniform expansion of electromagnetic radiation, light, then measures the same as we measure the movement of matter, or its failure to move, in the Earth's field of attraction.

By the late nineteenth century, it was widely realized that the Earth was radiating heat into space, and Kelvin used that fact to estimate the age of the Earth at between twenty and a hundred million years, probably as accurate as we'll ever get.

However, it was mistakenly assumed that Darwinian evolution would take billions not millions of years, and as a result, Kelvin's estimate was relegated to the graveyard of religious relics. With billions of years needed for evolution, the cooling Earth was forgotten and the meteorological notion of thermal equilibrium was adopted.

Under this incredible idea, the amount of heat the Earth receives each day is exactly equal to the amount it gives up, about as immeasurable an observation as any empiricist could come up with, which makes it immediately and totally acceptable.

No one can disprove what can't be measured.

As a result, when the thermal emissions radiating from the surface of the Earth were finally remeasured on an accurate basis by the use of radio receivers as noise and found to be consistent no matter in which direction the radio receiver was pointed, empirical science ignored the obvious source, right underfoot, the direction from which the receiver was always pointing, and instead declared the noise to be background radiation that was coming from every direction toward which the receiver was pointed, every direction being the great Outthere.

Once we decided that the noise was coming from outer space, it became the predicted fact needed by cosmology to prove the big bang theory and legitimize itself as a qualified member of the family of empirical sciences, one that can predict the existence of made-up facts that can then be found in reality. The noise that was produced by radio waves was the background radiation from the original explosion that brought space and time into existence, and produced the matter that occupies space and time.

When we discovered that the matter that occupies space and time wasn't as uniform as the background radiation, we sent a satellite up to prove that the background radiation was actually as irregular as the matter in space and time. Of course, the fact that the satellite was now several hundred miles above the atmosphere causing the low frequency radiation being emitted from the surface of the Earth to now measure as irregularly as the galaxies only provided more empirical proof that space and time, neither of which objectively exists in reality, really did come into existence in the type of explosion that normally takes things out of existence on Earth.

The fact is the Earth does emit. It doesn't emit light, which in any event would just bounce off the surface or penetrate the surface of matter. The Earth emits low frequency emissions which expand over the surface of an expanding sphere whose structure is made up of expanding flows of electrons being recombined by their associated inductive flows. These flows pass through the matter just as nonexpanding X-rays do, but because they are emissions, expanding as they do so, recombining in the process so that matter on the surface of the Earth is stuck to the surface of the Earth.

If you want definitive proof that matter moves in expanding flows of light toward the source of the light, examine the next light bulb you change. Empirical science explains the crud on its surface as the result of electrostatic attraction, based today on magnetic moment, yesterday on effluvium, but it's actually there as a result of the by-product of the mechanism that regulates the uniform expansion of the light being emitted at the surface of the bulb.

The mechanism that regulates the uniform expansion of light is also the attractive mechanism, the mechanism that produces the force that causes objects to fall to the surface of the Earth.

10. The Specter Of Spectra

Galileo sought to demonstrate the Copernican assertion that the Earth went around the sun by showing that the sun's apparent motion around the Earth was the result of the Earth's rotation. Because the Earth had no apparent cause for its rotation, a current force causing it to rotate was not apparent, Galileo sought to prove that a rotating object that had no apparent force *stopping* its rotation would continue to rotate forever. The force that Galileo dreamed up to keep the Earth rotating forever was inertia, and Galileo set about to prove the existence of inertia so that he could state that with space being empty, frictionless, inertia kept the Earth rotating on its axis and thus the sun's motion was only apparent.

A half a century before Galileo sought to prove inertia, Tartaglia had gone out on the proving grounds of the time and shot off cannonballs at various elevations to determine by trial and error the angle at which a cannonball would travel the furthest. Galileo mathematically measured the way balls would fall as a result of the momentum obtained from rolling down an inclined plane, and declared that, by mathematics alone, he could have predicted the outcome of Tartaglia's physical measurement in reality without knowing those results.

It was unimportant to Galileo that the inclined plane merely duplicated the angle of the cannon. What was important was the concept of inertia that Galileo injected into the measurements. He said that the ball moved forward after it reached the end of the inclined plane because it had inertia and in fact, if the ball had a frictionless path on which it could travel around the Earth, then it would travel around that frictionless path forever as a result of its inertia.

Therefore, in true empirical fashion, Galileo concluded that the idea of inertia was a fact because the motion of a ball on an inclined plane in the laboratory could predict the motion of a cannonball on the proving ground!

When it turned out that Copernicus was correct, that the Earth did travel around the sun, we were left with about two thousand years of delusional thinking, including about a hundred of those years in which the anti-Copernican forces, academics one and all, ardently argued that the sun, in fact, did travel around the Earth. As a result, much thought was given to how it is that we come to know reality.

How can we tell when something is, in fact, a fact, and not merely a delusion?

Galileo's mission in life was to prove that the Earth orbited the sun. To accomplish this, he had to prove that it was the Earth rotating on is axis in frictionless space that gave the sun the appearance of movement. Because God was not present spinning the Earth, he created the concept of inertia to explain why the Earth was spinning. Galileo was therefore faced with the problem of how to turn a concept, a hypothesis, something that he had made up out of whole cloth, into a fact.

Francis Bacon said, balderdash. Galileo's conclusion that the ball would roll forever on a frictionless path was beyond demonstration. How matter moves could only be measured when the matter moved, not how someone thought it might move. What made it move, on the other hand, was not measurable, was in fact, a secret. We can never know the secret motion of things, Bacon concluded, and specifically what makes the Earth rotate on its axis, unless we use our minds to analyze facts.

Galileo won the argument, and we ended up with a methodology that, because it was based on mathematical certainty, would forever protect us from making the same mistake the pre-Copernicans made.

How can we go wrong if we limit ourselves to measuring things?

Simply because measurements in and of themselves have no meaning.

It is our minds that fill in the blank spaces in the pictures of reality we construct out of our measurements, in Bacon's words, the secret motions of things. These blank spaces, what causes the Earth to rotate, what causes planets to orbit, what makes objects fall, what happens to light in a prism, can never be measured so that we can never know with certainty whether the pictures we devise to fill in the blanks spaces are in fact, facts, or just made-up stuff.

A methodology that allows us to delude ourselves into believing that the things we make up to fill in the blanks of the pictures we construct from our measurements are actually facts results in the proliferation of facts that do not exist in reality.

We can go wrong by limiting ourselves to measuring things because it is not the measurable facts we need to control, but the interpretation of those measurable facts, and these interpretations, these hypotheses, these made-up things can never be proven.

We start by deluding ourselves that we can turn our hypotheses into facts, then we delude ourselves by creating still additional hypotheses based on the facts we have created by the process and then we go out and delude ourselves into believing that we can prove those facts. Pretty soon we reach a point in which our reason, the only thing we have to analyze physical reality, the only tool we have to fill in the blanks, is discredited in the face of the bogus mathematical conclusions that result from crunching our made-up facts.

Nothing illustrates this better than the monkey see, monkey say conclusions dealing with the empirical observation of what happens to light when it passes into one face of a triangular slab of glass and comes out another, spreading out in specific colors. By assuming that spectral light is a window into the interior of the atom, we have created an atom from the endless series of proven hypotheses that is so complex that it can only be the product of the human mind. One of its features are electrons which whiz around a nucleus as a result of some unnamed force, and which, at the slightest provocation, change those orbits precisely in accordance with our mathematical conclusions. As the electrons change orbits, they create or absorb light waves.

Light becomes the occult result of an electron bouncing backing back and forth in orbits around the nucleus of an atom rather than the result of electrons themselves leaving orbits around the nucleus of an atom.

In contrast to this insight into the microcosmic, the part of physical reality that is too small to directly measure, we can move our spectral conclusions to the microcosm, the part of physical reality that is too large to measure. Viewing the spectrum from this mental vantage point, we can prove conclusively that all of the stars are rushing away from one another as a result of, in the tradition of Young's two-slit experiment where we measure what doesn't exist, a shift in a part of the spectrum that doesn't appear in our detectors, absorption lines, that is analogized to the shift in the pitch of sound waves as they are propagated through air from a moving source.

The further the source, goes the reasoning, the greater the absence of light shifts toward red, and the faster the source is receding.

This leads in turn to the once ineffable conclusion that the stars were all together at one time and this gives us, indeed, a lot to measure, including when they were put in motion by a big explosion, that, once blessed by the Pope as being consonant with the concept of God and creation, could take its place among all the other delusional scientific facts as worthy of generating additional provable hypotheses.

What exactly is happening when light enters one face of a prism as white light and emerges from another as color in lower frequencies?

We know a number of things that are facts. We know that the light that is entering the face of the prism is expanding uniformly over the surface of a sphere that is diminishing inversely with the square of the distance from its source. We know that the light is capable of being divided when it encounters glass. If we are on a train traveling through a tunnel, we can see ourselves in the glass. If we exit the tunnel, we can no longer see ourselves. If the train plows into the night, we can once again see ourselves in the glass, and outside observers can see us just as clearly, more sharply, in fact.

Outside observers can see us more sharply because the image that they see has become sharper, more defined, as a result of passing through the glass. The light striking the object on the light side of a glass is expanding in expanding spheres from the various points that make up the object. When the light encounters the surface of the glass, the expanding spheres divide as a result of reflection, with some of the flows bouncing back into the lighted area of the car, where they are washed out if there is a source of light behind the glass, and the rest passing into the glass, where a certain percentage is dissipated as heat.

The remainder of the light passes into the area outside the glass. After the flows pass through the glass, they continue to expand over the surface of their expanding spheres. However, there is less light to expand over those surfaces. As a result, the recombining flows of light leave gaps, dark lines, between the various flows, lines of no light that wouldn't have existed had the flows not been diminished by the glass.

When expanding flows of light pass through glass and the remaining flows are recombined, areas of no light are left between contiguous expanding flows which delineate the points of the object off of which the light is expanding. We see the object much more sharply. The same effect can be seen by looking at objects through a pinhole. The amount of light making up the expanding flows on the eye-side of the pinhole has been vastly reduced, and when it is recombined, the areas out of which the light has expanded delineate the object so that it can be seen with much greater definition.

We want to concentrate on what is happening to light as it passes through the glass between the point of entry and point of exit, because we know that at the point of exit in a prism, the light emerges in different frequencies which, because only like frequencies can recombine, recombine into expanding flows of separately colored light.

Light can be conducted for the same reason that

electricity can be conducted: There are spaces between the atoms that make up the conductor which are relatively free of orbiting electrons and which therefore allow the flows of electrons to pass freely. There are, however, distinct differences between how light and electricity interact with the matter that conducts them as well as between the matter itself.

While electricity can be produced as alternating current in which the current direction changes continuously and this change in current produces an oscillation which might be mentally confused with frequency in light, a flow of electricity is a flow of electrons without frequency. Electricity is also a nonexpanding flow, merely following a path between an excess of electrons and a deficit of electrons as a result of the excess affinity propensity at the area of electron deficiency. Light, on the other hand, has frequency and expands when it interacts with the matter that makes up its conductor.

Further, electrons have to be injected into the conductor while light is coherently structured by its inductances before it enters the conductor. As a result, with electricity, the conductor organizes the flow of electrons into electricity while in light the conductor disrupts the flow of electrons that make up the light.

In addition, the electrical conductor contains nuclei of atoms with orbiting clouds which are replaced when the current begins to flow and which then become a part of the current flow, causing a delay in the establishment of a current flow when the circuit is completed and a lag in the cessation of current flow when the circuit is broken. Because electrical flows are not themselves structured, the alignment of the atoms of the conductor itself is not critical to its ability to conduct the current with the flows capable of establishing and maintaining paths in accordance with the existing affinity propensities within the conductor. Light conductors, on the other hand, are composed of nuclei positioned into a web of stable equilibrium.

In dealing with an electron with its two properties of at rest motion and affinity propensity, there are only a limited number of configurations into which it can form. One of those configurations is the unit in the nucleus of an atom in which the electrons come together as a result of their affinity propensities overcoming their at rest motion. Another structure the electrons can form are expanding spheres of light in which the excess affinity propensities of the flows of electrons moving at their at rest speed of light are balanced by their associated inductive flows.

The only other structure the electrons can form is a structure in which the opposing forces of at rest motion and affinity propensity lock them into a structure of stable equilibrium. This structure can be visualized by placing ten dots across the top of a page of paper, ten dots down the side and then filling in the remaining nine rows so that there are a hundred evenly spaced dots. If each of the dots have their at rest motion and affinity propensity balanced by the at rest motion and affinity propensities of the contiguous dots, then the dots will be locked into positions of stable equilibrium. If we cube the square by adding ten layers of one hundred dots, then we have a three dimensional structure made up of individual electrons each separated from the other and held together by the affinity propensities of its electrons.

This third structure is essentially the structure we picture that has matter made up of atoms immobilized by their positive and negative charges, but without the duplicative quality of repulsion.

In constructing an atom, we had three levels at which matter could evidence the type of excess affinity propensity that would produce stable webs of equilibrium. The first is the electron itself, and the structure made up of electrons chemically placed into positions of stable equilibrium by the organic matter of the brain generates the mind.

The second configuration, stable webs of equilibrium made up of individual units of matter, can't exist in an area of matter in which we exist because the units immediately combine with existing nuclei of atoms in the environment.

The third web of stable equilibrium is formed by the nuclei themselves when they have been stripped of their orbiting clouds of electrons. In light conductors, primarily glass, matter whose crystal structure has been altered by a process that strips the electrons off of the nuclei of the atoms that make it up and then places the atoms into positions in which their excess affinity propensities immobilize them in a web of stable equilibrium, the space between the nuclei allow for the relatively unhindered passage of light. In the mind, the at rest motion of the electrons has not been overcome so that the resulting structure is in a state of dynamic equilibrium, capable of being affected by passing affinity propensities, while in the light conductor, with the at rest motion of the electrons immobilized in the individual nuclei, the resulting structure is physical matter. A light conductor, then, conducts light to the extent that the nuclei of the atoms that make it up have been stripped of their orbiting electrons in a manner that allows the nuclei to form together into a structure of stable equilibrium with a minimal of orbiting electrons.

Light conductors can have different effects on light. Glass allows light to pass relatively unscathed. A prism, because it varies the distance the light travels, separates out the high and low frequencies in accordance with the distances that they have moved through the prism. Monochromatic filters, a misnomer because all light is a single frequency, alter the frequency into specific colors. Special crystals polarize light, only allowing it to expand over partial expanding spheres as if it had reflected off of an opaque surface, and while empirical science uses polarization as proof that light is a wave made up of electric and magnetic forces acting at right angles, polarization is merely the result of light being conducted by specific constructions of crystals.

The final difference between electricity and light and their respective conductors is that the electric flows in a conductor are massive when compared to the structured flows accommodated by conductors of light.

Unless matter is one hundred percent reflective, it has a physical boundary which can, to one extent or another, be penetrated by light. If we have a piece of tin and place it in the sun, expanding flows of light strike the surface of the tin. The sunlight being high frequency, but extremely low intensity flows of light, as opposed to the low frequency emissions from the Earth, its flows begin to replace the clouds of orbiting electrons holding the nuclei at the surface of the tin together. As these electrons are emitted as heat, along with the flows of light bouncing off the surface, flows of light begin to enter the surface. As the flows penetrate further between the atoms, they replace more electrons, but at the same time they begin themselves to break down as they are disorganized among the various excess affinity propensities that are created and now are competing with one another as a result of the flows of light striking the physical structure of the tin.

If the amount of sunlight striking the surface of the tin is insufficient to replace the affinity propensities of the nuclei making up the surface, the surface will maintain itself as it reaches a point of equilibrium with the incoming sunlight. If the sunlight penetrates the surface, the flows begin to break down into individual electrons and the individual electrons begin to reorganize themselves in accordance with the existing affinity propensities so that they can exit the surface as heat. If the tin is thin enough, some of the electrons might make it through to the other side and exit that side as heat.

So long as the matter remains intact, the equilibrium of the matter with the incoming flows will cause the matter to re-emit the flows, producing a change in frequency in accordance with the nature of the matter and the intensity and frequency of the flow with which the matter is interacting.

With glass, the fabricated light conductor, some of the light bounces off the surface, some is disorganized passing through the surface, and the rest emerges from the other side of the surface in flows that can be reorganized to approximate the flows that entered the surface. When the light strikes the surface of glass, it is diminishing uniformly over the surface of an expanding sphere, with its inductances balancing the presence of its flows. The light that bounces off the surface begins to expand from new centers of expansion and, depending upon the reflectivity of the surface and the angle at which the expanding flows strike the surface will contain an exact percentage of the information that the light contains. Depending again upon the reflectivity of the surface and the angle at which the light interacts with the surface, glass allows a substantial portion of the flows to pass between the nuclei of the atoms that make up the surface. However, the expanding flows of light penetrating the surface have to align themselves in new centers of expansion in accordance with the structure of the surface, bending in the measurable refractive index of the matter being penetrated.

As soon as the light reorients itself as it passes the

plane of the surface, it begins to expand over the surface of an expanding sphere once again. The flows will begin to lose some of the electrons that make them up, and these electrons will be emitted from the surfaces of the glass as heat just as happens in the example of the tin. When the light exits the opposite side of the glass, it readjusts itself as a result of passing once again into a different medium, bending in accordance with the medium's refractive index.

When the light enters the glass, it is moving at its at rest speed as moderated by the medium in which it is moving, the atmosphere. When it is within the glass, the light's speed is slowed by the affinity propensities holding the nuclei of the glass together. When it exits the glass, it resumes its at rest speed, as moderated by the medium into which it exits, the atmosphere. The extent to which it is bent upon entering and exiting the glass has nothing to do with its speed, which is merely the product of its at rest motion with relation to the existing affinity propensities retarding its motion in the glass.

As the light begins to re-expand within the glass of the prism, it is both slowing down and losing electrons from its packets. When light expands over the surface of an expanding sphere without interacting with matter, it diminishes uniformly in intensity but not frequency. The packets of light are splitting apart and recombining uniformly. The packets do not lose electrons, the light maintains its speed and thus the frequency of the light is not altered.

However, when the light expands as it passes through

the glass of the prism, it is losing electrons out of the recombining packets while it is being slowed down by the affinity propensities of the glass's nuclei, and this causes the frequency to lower. The further the light moves through the glass, the lower its frequency becomes.

Where the light exists the prism determines the distance it travels through the prism. After the light refracts and begins to re-expand over the surface of an expanding sphere, the point at which the expanding sphere touches the exit side of the prism determines the center of the expanding sphere and thus the shortest distance the light travels through the prism.

The light that exits the prism can only be recombined by frequency, and thus recombines as colored light. Because there is substantially less light left after it has been split at the surface and after some of the electrons that make it up have been lost as heat, the recombination process has an effect similar to the effect in the example of the light shining through a train window except that with prismatic light there is no object to define and thus nothing to define the areas out of which the light is recombined. Depending on the source of the light, its frequency and intensity, when the light recombines upon exiting a prism, it leaves bands of no light, areas of no color out of which the light has been recombined.

The shortest distance that the light can travel will result in a visible frequency of yellow light with the light expanding on either side of it having its frequency lowered in accordance with the distance it travels through the glass.

The second shortest distance is the path the light travels that is expanding toward the point of the prism defined by the two opposing sides through which the light is passing, and it is this distance that reduces the frequency of the white light to red.

Thus, the high frequency range, yellow to red light, recombines at the top of the prism, with red light moving the longer distance through the prism.

The third shortest distance is the path the light travels that is expanding on the base side of the prism beneath the white light whose frequency is being reduced to yellow light, and it is this distance that reduces the frequency of the white light to green light.

It is clear why red and green light have similar frequencies. The difference in frequency is the result of the slightly longer distance the green light has to travel before it emerges from the prism and this slightly longer distance is due to the path being closer to the base than the path the red light travels.

The lowest frequency light emerging from the prism is blue light, which is the result of the white light traveling the longest distance it can travel through the glass of the prism before it simply dissipates into individual electrons and exits as heat.

Thus, the low frequency range, blue to green light, recombines at the base of the prism, with blue light moving the longer distance through the prism.

What the prism does is break light into the two frequency ranges that the eye perceives.

The eye does not see a continuous spectrum between blue as the lowest frequency and yellow as the highest frequency. It sees two ranges of frequency, the low range which stretches from green to blue, and the high range which stretches from yellow to red. This is why these four colors are basic, they anchor the two ranges of the eye's frequency reception, as evidenced by the eye's rods and cones. When the eye sees reflected color, it see a single frequency at any point in either range. When it adds light colors, it sees more than one frequency which together add up to less than the frequency of yellow light.

Light refraction in a prism merely breaks light out in these two ranges, delineating the high range, yellow to red light in the shorter distance between the opposing sides, and low range green to blue light in the longer distance between the opposing sides, with the distance determining the resulting frequency.

We see this same distribution of color when the sun strikes the Earth at the angle that produces colorful sunrises and sunsets. Sunrises and sunsets are the book ends of a blue sky, and the most visible display of ignorance involving misplacing the blue at the wrong end of the visible frequency spectrum can be seen most days by simply looking up at the sky.

Why do we see a blue sky?

Not only do we see a blue sky, we see a yellow sun.

What happened to all the reds and the greens? If we look at light through smoke, so the story goes, the particles of smoke scatter the light so that the light takes on a bluish tinge. Here indeed is an empirical truth on which to anchor all of our understanding about light. Something in the air is scattering the white light so that the only thing we see is blue light. Because the things in the sky are called clouds, and they aren't blue, it must be the molecules of air themselves that are causing the light to scatter.

Of course, we have no idea what this scientific process of scattering involves and under the classical concept of light, in which the incoming sunlight contains all colors, this scattering would involve having all the colors but blue being absorbed. However, other than these minor problems, ask an empiricist why the sky is blue, and he'll tell you its because the molecules of air are scattering the light. Ask him why it turns red when the sun goes down, and he'll tell you because the light is scattered more.

Blue light, which is the light refracted from passing through the longest portion of the prism, is the lowest visible frequency. In the daytime sky, it is breaking out on the inside of both natural prisms that simulate sunrise and the sunset.

The incoming sunlight strikes the atmosphere at two angles which represent the sunrise and the sunset and which equal the angle that light enters a prism. In both the sunrise and sunset, the sunlight that passes through the thickest portion of the atmosphere, toward the day side, refracts blue light. The white light that breaks out on the night side is passing through the thinnest portion of the atmosphere and breaks out as red and yellow light.

If we are sitting outside at high noon, the sunlight is making an angled entry to our east and our west, our east being the sunset for people over the eastern horizon and our west being the sunrise for people over the western horizon. The two prismatic effects are mirror images of each other, with the blue frequency breaking toward the day side of the planet, the place we are sitting at high noon. The sun overhead is not participating in the prismatic effect, its light being reduced directly to yellow light by the atmosphere, and the yellow light is what is bouncing off matter in physical reality to produce the colors we see all around us.

Over each horizon the yellow light is in the direct line of sight of the sun with the red spreading out around it depending on the position of the sun. This is true whether the horizon is viewing the sunset or the sunrise. There is one difference at sunset, however. Because the sunset is occurring over warm air masses which don't exist in the area over which sunrise occurs, the air masses are more moist. As a result, the red, which is not a frequency above yellow, but rather a frequency below yellow and above green, is frequently reduced to green light just as the sun is sinking beneath the horizon.

Astronomers can use a prism to produce spectral colors from starlight or they can use the etchings called diffraction grating. Diffraction grating produces a cacophony of colors as the light is refracted and recombined. Viewing these recombined flows from separate directions with optically enhanced instruments allows the construction of a continuous spectrum of starlight that, when the individual observations are arranged in a strictly empirical fashion, produces the spectrum seen in a prism.

A strictly empirical fashion, of course, means arranging the light collected from the diffraction grating the way light emerges from a prism.

When we pass light through a prism, what we see is what we get, the frequencies that the glass that makes up the prism can conduct, yellow light, the highest frequency, in the center, red and green light, lower frequencies, to either side and blue light, the lowest frequency on the bottom of green light. This corresponds with the way frequencies arrange colors with respect to luminosity, color adaptation and color adding, with yellow light being the highest frequency, and red, green and blue light becoming progressively lower frequencies. In color adding, color as the result of frequency is clear, with red and green light adding to produce yellow light, which, when the lowest frequency, blue, is added, becomes the original light that entered the prism, white light.

A white light source in the laboratory is sufficient to provide enough light so that a continuous spectrum is produced when the light is passed through a prism. The less light collected by a prism, or by instruments using a diffraction grating, the less light there is to recombine.

When there is less light to recombine, light recombines out of areas leaving no light in areas between the recombined light.

What do we conclude when some of the light doesn't come out of the prism or isn't available for diffraction grating to make up a continuous spectrum?

First, Newton used the fact that a prism turned white light into colored light to reverse reality by assuming white light contained all colors, forcing us in our Newtonian adulation to believe that matter absorbed all colors, reflecting only one.

Light from a strong source emerges from the prism so that one frequency shades into the next. All of the colors that make up the white light under the Newtonian concept are therefore present in the white light.

Then Young, by concluding that the absence of light was actually the presence of twice as much light, locked mindless empirical devotees into thinking that when we saw nothing, we were actually seeing something. We were precluded from seeing that light was being recombined so that when we saw no light on the collection screen, that was exactly what was on the collection screen, no light.

Combining these two misassumptions, empirical science was in for a shock when it found that starlight that was passed through a pinhole, refracted with a diffraction grating, recombined into individual colors and then distributed in the order that light would come out of a prism had it come out of a prism, was missing some light when it was recombined.

Knowing that expanding flows of light recombine, and that when some of the light has been removed from these expanding flows, that gaps are left in the light, areas of no light out of which the light has been recombined, just like the bars of no light in Young's two-slit experiment, we would be content with finding this effect, and not attach much importance to it.

However, having been raised in the belief that white light contains all colors, and that the light that we had beat to death in our laboratory, overexpanding it with the pinhole, fracturing it with diffraction grating, and then arbitrarily assembling it like the prism orders the frequencies it produces from white light, we look for what isn't there because we knew it was there at one time.

In the nineteenth century, bright-line spectra were discovered. As we have seen, frequency is the result of the rate at which atoms combust. Because all elements are unique in the size of their orbital clouds, the frequency or frequencies that a combusting element produces is unique. Thus, each element undergoing combustion produces a particular frequency of color, a bright line of color, and because the spectrum produced by a prism produces all colors, that bright-line frequency can be placed on the spectrum of colors to give each element a spectral fingerprint.

The fact that there are dark lines in the reconstructed spectrum of stars, and the fact that each element produces a particular frequency when it combusts should not surprise us. What should surprise us, what should in fact amaze us, however, is what empirical science can do with these two facts.

Well, maybe not.

When empirical science looks at the missing bands of light in the spectrum reconstructed from starlight, it says that because the color was once there, it has been absorbed out of the spectrum after it was emitted on the surface of the star. Because we know that all elements have a bright-line spectrum, all empirical science has to do is line up the bright-line spectrum on the absorption spectrum and, *voilà*, it can say that because the element that produced the bright-line spectrum falls more or less on an absorption line, that's proof that the element is on the surface of the sun.

The element has to be on the sun, and it has to be emitting its spectral fingerprint because its spectral fingerprint doesn't appear in the spectrum of light produced by the star. White light contains all colors so that if one of the colors is not there, it must have been absorbed out of the light.

What could have absorbed the spectral fingerprint out of the star's spectrum?

Empirical science views light as a wave phenomena. As a wave, light is not something that is emitted from the atom, but rather is a result of something that is occurring within the atom, the shifting energy levels of the discrete orbital electrons within the atom itself. When an orbital electron changes energy level, a light wave is produced. Even though the light wave produces a bright-line spectrum, the waves of starlight contain all colors. Thus, when starlight containing all colors passes through atoms of a specific element, the electrons that would produce the bright-line spectrum absorb the wavelength instead, removing it as one of the total colors being transported in the white light.

It is the assumption that each element absorbs its own wavelength out of white light which contains all wavelengths that is at the basis of the empirical claim that we can know what the elements are on the surface of the sun and other stars, and in the vast clouds of elements drifting in interstellar space.

Whenever the absence of light shows up as a dark line in the absorption spectrum, we see the bright line of light, the frequency emitted by a particular element, and we know that the element is there.

Instead of seeing light where there is no light, we see matter!

For instance, the bright-line spectra produced by burning calcium, sodium and hydrogen are pasted in a continuous spectrum from a prism along with the dark lines of no light created by spectral analysis of starlight to find calcium, sodium and hydrogen on the surface of stars.

Once pasted in their empirical places, these bright-line spectra lead to the red shift, the analogy of the red shift to sound waves, the conclusion that stars are all moving apart one from the other, the corollary to that conclusion, that if they are all moving apart, they were all once together, and the necessary corollary to that corollary, now a conclusion, that something that was once together and is now burning and flying apart has to be burning and flying apart as a result of an original explosion.

With Young's two-slit experiment, we looked at the absence of light and saw light.

With an absorption spectrum, we are looking at absence of light and seeing matter.

Here we are looking at the absence of light shifting and seeing the beginning of the universe!

The Big Bang, blessed by the Pope as an acceptable creation myth, displays empirical science in all of its glory, meticulously measuring what doesn't exist, in this case, the missing light from a spectrum or its engineered equivalent, and coming up with conclusions closely supported by objective observation. We look at a rainbow and we see matter being created out of the type of explosion that would obliterate it in our laboratories. The pot of gold at the end of the rainbow turns out to contain the secrets of the beginning of the universe.

The movement of matter in space does not, of course, affect the frequency of the light the matter is emitting.

When we bounce frequencies off of an object moving in space, as orbit determination teams do when they are trying to locate where our space probes are, or off of moving air masses as weather forecasters do, we can obtain information from the manner in which the movement of the object alters the frequency. While this effect is referred to as the Doppler effect, the analogy to sound waves is inappropriate because, the wave analogy aside, estimating distance by the change in frequency of signals bounced off a moving object has nothing to do with determining the speed and direction of a moving object by the frequency the moving object itself emits. The ability to estimate is merely the result of a frequency change that results from the original frequency interacting with moving matter.

However, the eye, or a detector substituting for an eye, could see changes in frequency if they moved with respect to an emitted frequency. If we were moving toward a star with a high enough velocity, then more frequencies per unit of time would pass the plane of the eye's detector, and the frequency would appear to be higher. It would, in short, move toward the red or yellow end of the spectrum.

If the eye or its substitute detector were moving in the same direction that the light was moving, on the other hand, the frequency would appear to be lower, move toward the green or blue end of the spectrum. If we were moving away from a star with a high enough velocity, we would see the light as more blue than it would otherwise be because, the faster the eye moved in the direction the light was moving, the slower the frequencies would pass the plane of the eye.

The ability of the eye, or its substitute, to detect a change in frequency as a result of moving opposite directions in relation to a fixed star located on the equatorial plane of the sun has led to one of the more amusing misperceptions in modern astronomy. *At the Gates of the Citadel*, the first volume of The Copernican Series, describes the failure of Kepler's Laws, upon which Newtonian Mechanics is based, to describe reality. Empirical science, with its head up the back end of its detectors, forgot that the movement of the Earth in relation to the sun can be measured with a stick planted in the ground and a little patience.

The result shows that the time differential between Kepler's equal points is three days rather than the equal times required by Kepler's Laws.

Correcting Kepler's equal points to their actual positions in reality reduces the time to a little over a day and a half. Because we have an object, the Earth, traveling around the same apparent distances of a moving focal point, the sun, we can compute the speed of the sun using these time differentials.

More important, we can determine with absolute precision the direction in which the sun is moving!

Picking a star in the approximate direction of the sun's movement gives us Betelgeuse in the Orion spur.

Betelgeuse happens to be one hundred and eighty degrees away from Sagittarius, the star toward which the stargazers, basing their measurements on empirical science, have the sun moving, one hundred and eighty degrees being the pivotal percentage describing the gap between the perception of physical reality by empirical science and physical reality itself.

Why do the stargazers get it backward?

Because empirical science has the blue and red ends of the spectrum reversed. When the stargazers measure blue, they think they are moving toward the star. When they see red, they think they are moving away from the star.

In fact, a shift to blue shows that we are moving away from the star and a shift to red shows we are moving toward the star.

Measuring the spectrum of Sagittarius as moving toward the blue shows we are not moving toward it, it shows we are moving away from it.

The red/blue reversal reverses the way empirical science sees the sun moving in space!

It also demonstrates the correctness of the computations as to the sun's speed and direction and thus the failure of Kepler's Laws to describe reality.

When the Royal Society of London demanded that its employee, Robert Hooke, find a stellar parallax to prove the Copernican proposition in the seventeenth century, Hooke, following the demands of his patron, found stellar parallax.

Stellar parallax, the shift of the position of a star against the stellar background as a result of measurements taken when the Earth is at points one hundred and eighty degrees apart in its orbit, requires a lot of corrective computation to account for refraction, aberration, hopefully the precession of the equinoxes, and the direction in which the sun is moving. Hooke's measurements of the parallax of a star in the Constellation of Draco turned out, like Hershel's measurements demonstrating his discovery of Uranus, to be one hundred and eighty degrees wrong.

Parallax measurements are used as the basis of all estimates of the size of the universe. They are so tenuous as to leverage minor errors into vast distances. Hooke's error was harmless and put food on his table. Whether computing parallax measurements with the sun moving in the wrong direction is equally harmless is questionable. Hooke was just attempting to show that Copernicus was correct, that man wasn't the center of the universe. Parallax measurements, on the other hand, produce a universe so immense that not only is man insignificant, he is alone, which I guess on a philosophical basis means that if we're not the center of something, eliminating everything else is a close second.

And the mental state of mankind with respect to the immensity of the universe, this noble isolation, is produced simply by mixing up the ends of the spectrum!

In addition, confusing the direction the sun is moving by the frequency shift of a star's light as a result of the movement of the Earth around the sun won't tell us anything about the motion of the stars with respect to the

Earth. Frequency itself does not contain information about velocity or direction.

Nor does the light emitted from stars contain all of the colors of the spectrum. The diffraction of starlight and its arrangement into a sequence that mimics the way light emerges from a prism is merely a meaningless process. Starlight is composed of a single frequency, and there is not a red component within that frequency toward which absorption lines are capable of shifting. The reduction of starlight at various frequencies by overexpanding it with a pinhole and then refracting it with a diffraction grating merely distributes the recombined starlight across the spectrum in accordance with its original intensity, intensity being how much starlight there is available to have its frequency reduced in the process.

Nor is light analogous to sound waves. Sound waves are an actual disturbance, some sort of vibration in an actual medium, the atmosphere as held to the surface of the Earth by the attractive force. These sound waves actually do vary with the source of the sound, the Doppler effect. If a train is standing still in the night, the whistle will maintain a constant shrill. If the train is receding into the night at two hundred miles per hour, then, with respect to a particular ear, the number of times those waves wash over the ear per unit of time is diminished, just as it is increased with respect to the ear toward which the train is racing at two hundred miles per hour. The sound waves are an actual physical disturbance of a medium, the disturbance exists at any point within the medium, and the movement of the source affects how those waves are generated in the medium, and the positioning of the receptor determines how those generated waves will be perceived.

Light has no medium to disturb, its nonexistent medium has no force holding it in place, and because light is a release of electrons at their at rest speed, the motion of the emitter has nothing to do with the speed at which the electrons are released or the frequency that the released electrons produce.

Nor does a displacement of the absorption lines, bright-line spectra of elements such as calcium, sodium and hydrogen in the spectra of starlight toward the red side of the prismatic spectrum signify that the star is analogous to a receding train whose sound waves, becoming longer, can identify which way the train is traveling. Arranging the spectrum so that red light is the longest wavelength reverses the direction the absorption lines would shift if the absorption lines were to be shifting toward the longer wavelength. Blue is the lowest frequency, not red the longest wavelength, so that a shift of the absorption lines to the blue side of the spectrum would be required for stars to all be moving apart.

If starlight contained all colors in the spectrum.

If starlight were analogous to sound waves.

If starlight were affected by the movement of the star to begin with.

We're seeing something, though, so what is it? When we take starlight and correlate it with the apparent brightness of the starlight's source, we find that the dimmer the starlight, the more the areas out of which light has recombined appear to move into the red areas of the spectrum.

We only have to look at how light moves to see what is happening. Light diminishes inversely with the square of the distance over which it expands, which in simple terms means that the further the star, the dimmer the light, the dimmer the light, the less light there is, and the less light there is, the less light there is to be diffracted.

Instead of the absorption lines moving toward the red end of the spectrum, light from the blue/green range, with the lowest frequency, is dropping off the visual range of perception as the intensity of the starlight diminishes inversely with the square of its distance. At the same time, light from the red/yellow range is also diminishing.

More yellow light is diffracting out as red light.

The absorption lines are the result of less light recombining after refracting. They aren't shifting toward the red, rather, as the starlight's intensity lessens, the entire range of diffracted frequencies is moving toward the blue, or lower frequency side of the spectrum.

The absorption lines are stationary. They are pasted in place by assuming their location from the bright-line spectra.

It's the spectrum that's shifting!

Which means that the red shift evidence for the expanding universe is evidence only of the ability of empirical science to produce delusional explanations for nothing as fact.

And to do so in a manner that almost exactly reverses reality.

We've already seen that the evidence for the big bang, the logical starting point for an expanding universe, is delusional. Instead of sourcing the three percent static, picked up uniformly no matter which way radio receivers are pointed, from the ground beneath the radio receiver, empirical science sources it as cosmic rays from space and labels it as the remnant of the big bang explosion.

When the distribution of the galaxies in space was found to be nonuniform, empirical science simply moved its detectors several hundred miles above the surface of the Earth where the static from the Earth's emission field would no longer be uniform, and declared that the static, being as nonuniform as the galaxies' distribution, doubly proved the big bang.

Every misassumption ever conceived about light comes together to produce a certainty of knowledge that exceeds empirical expectations, knowledge which, because light doesn't exist between its source and its destination other than as a disturbance of a medium that the Michelson and Morley experiment proved didn't exist, can never be checked, and uncheckable, never disproven.

This twentieth century concept of light and energy will be but a footnote in the history of the acquisition of knowledge stating that quantum mechanics was simply an occult view of reality in which reality was merely a reflection of the observer's fantasies.

For the time being, however, the belief that light is not made up of what gives it off precludes comprehending just about everything we want to understand about physical reality and the world in which we live.

This delusional view of what's going on in reality extends to our everyday life.

Just as locking gravity out of reach as a property of matter removes it from rational analysis, locating the cause and effect of light in a microcosmic mishmash of cause and effect with no physical processes in between leaves us at a loss to explain the most obvious things in our existence, why flowers open up in the morning sunlight, what lightning is, why tornadoes trail thunderstorms.

In the monkey see, monkey say world of empirical science, what monkey says becomes what monkey sees, and we end up seeing what doesn't exist and not seeing what's right in front of our eyes.

11. Sunrise To Sunset

We can only know the microcosm by its effects. We play around with physical effects and find that bombarding matter with something produces sparks in gas, paths in clouds, marks on film, electrical impulses in wire meshes. Sparks and marks have number, location and intensity, paths curve and spiral, electrical impulses curve and spiral in numbers and intensities. We can't see what is causing the effects, and we don't know what is producing the cause. The only way we can "see" what is going on is to make up pictures in our minds.

But empirical science, created to avoid the formation of inaccurate pictures in the wake of the greatest inaccurate picture imaginable, the picture of the sun going around the Earth, doesn't have a method to ensure that the pictures we form have any relation to the physical reality we measure.

Empirical science claims that mathematics can prevent us from falling prey to colossal blunders by doing our thinking for us, telling us that if we form a hypothesis about the sparks and paths we see, and that hypothesis predicts the existence of other measurable sparks and paths, then when we measure the predicted sparks and paths, we can accept the hypothesis doing the predicting as fact.

All of the sparks and paths are caused by electrical and magnetic phenomena, yet we don't have any picture at all as to what these phenomena might be. In fact, we haven't the foggiest notion of what produces them, how they work, or why they produce the effects they produce. We're just sitting around monkey see, monkey saying, making up stuff as we go along, but saying that, because the stuff we made up turned out to account for a fact we didn't know existed when we made the stuff up, that the stuff we made up is real.

When we look at the motion of matter in space, we assume straight-line motion and then make up matter to change that motion. When we look at the curves and paths in our cloud chambers and particle detectors, we see everything but straight-line motion, ascribing the curves and paths to an increasingly complex array of particles containing different mixes of the old standbys, mass, charge, magnetic moment, things for which we have no explanation other than the things that they explain.

We see straight-line force in the macrocosm, where it doesn't exist, and ignore it in the microcosm, where it does exist.

If someone attempts to provide a picture that actually explains the phenomena that we measure, the picture inevitably opposes the concepts that have been mistaken for facts. Concepts that oppose scientific facts, the stuff we made up and now believe is real, are nothing, no matter how well they explain actual facts in reality.

In short, we have made up a bunch of stuff about how we think reality should be, we have discovered methods to manipulate that reality, methods that are limited by the stuff we made up and for which we have absolutely no rational explanation, and then we have used the stuff we have made up to see the cause of the effects of those manipulations, totally blinding ourselves to both the physical reality we see and excluding ourselves from any access to an actual understanding of how that physical reality works. It would really be quite frightening to see empirical science, wallowing in absurdities, violently strike out against rational explanation in an effort to protect its occult practices if the picture of highly trained and intelligent minds blindly bumping into physical reality through the mists of their incomprehension weren't so amusing.

And the amusement would be more than hollow if it weren't for the fact that empirical practices have left us not only ignorant of physical reality, what we are, and what our place in the universe is, it is liable to leave us that way to the bitter end, until we die on the face of this planet, never having ventured out to take our place in the universe.

Not trusting ourselves to think, we have put our faith in our measurements to think for us, refusing to admit that our measurements require thinking to be meaningful. As a result, we see all sorts of particles, alpha particles, beta particles, gamma rays, we see rays come from the cosmos, we see solar winds blowing the sparkling tails of ice balls in space and then bury tanks deep in mine shafts to detect the beginning of time, send up satellites to detect the beginning of space and build circular electromagnetic race tracks miles wide to detect the building blocks of time and space. We make up stuff to detect, make-up detectors to detect the stuff we make up, and when we finally fabricate detectors that will detect the made-up stuff, we treat the made-up stuff as fact and use it to make up more stuff that needs even more sophisticated detectors to detect.

Empirical science lays no claim to making sense out of what it measures, so what it measures doesn't have to make sense. It is a never ending process in which nothing is unacceptable so long as it doesn't disagree with what has been accepted.

And all the time the sun is sitting in the middle of the solar system, emitting light in an expanding sphere of such intensity that by the time it reaches the Earth some ninety million miles away it is still sufficient to produce over a thousand watts of power per square yard of Earth it strikes measured by the sheer pressure it exerts when it bounces off of a calibrated mirror. And because empirical science has no way to deal with an expanding sphere of light on a quantitative basis, it measures it in terms of its heat as expressed as energy, and doesn't even see the light itself.

Light illuminates the universe, it is what we rise to each day, it is what we see by, it is the only way we know that physical reality exists beyond the realm of our touch, and empirical science just says there's nothing there, it is a local effect that travels from one place to another through a medium that can only be structured with chalk on blackboards snuggled away in quite buildings buried in verdant splendor.

The fact is, the Earth is a ball of matter cooling in space from its original state of being a miniature star made up of the most complex atoms possible unraveling as a result of field replacement in the stronger emission field of the sun. The cooling process, quite normal for burning matter surrounded by a vacuum retaining no temperature other than the light moving through it, eventually permitted the atoms of matter at the surface of the Earth to stop breaking down into their constituent nuclei, units, and electrons, and begin to break down into other matter, matter capable of maintaining its substance as the white hot sphere of combustion receded further beneath the forming crust.

As the Earth crusted over, other atoms, having broken down into less solid atoms, were captured in the fierce attractive mechanism which is the by-product of the mechanism that regulates the uniform expansion of the Earth's expanding sphere of emissions, and formed a thick gaseous cloud of light elements not unlike what we see on Venus and Jupiter. As the cooling process continued, these elements began to combine and precipitate out, the most common combination being hydrogen and oxygen to form lakes and oceans.

All the while the Earth was cooling and forming oceans between its crustal masses, it was spinning in front of a sun that caused its surface to alternate between night and day. Everything that happened on Earth during its formation, and everything that happens on Earth today, as it cools down and begins to die, happens as a result of the planet rotating on its axis so that it experiences cycles of sun and no sun, light and no light; and empirical science, incapable of seeing light, let alone seeing it for what it is, has no way to connect all that happens in front of its very eyes to reality, opting instead for occult explanations, and failing even that, willful ignorance of the facts and blind attacks to preserve its process.

From the very moment the first rays of the sun strike the margin between night and day, things begin to happen on the sunny side of the margin while, across the face of the Earth, at the other margin, things cease to happen as the Earth passes out of line of the light coming from the sun.

The profile of the expanding sphere of light produced by the sun is of a sphere with a stronger emission field toward the equatorial region of the sun because the sun's rotation concentrates its molten surface to some degree at the equator at the expense of the poles. The Earth as well as all of the planets and moons occupies this stronger field as it orbits the sun, the interaction of its own emission field with the sun's emission field being the cause of rotation, and the reverse bias of the mechanism of attraction embedded in the sun's expanding emission field flowing out and back from the direction of rotation being the cause of orbiting.

The sunlight is always there, expanding out away from the sun. It is the Earth that is rotating in front of the sun and it is the Earth that has to adjust to the effects of the light as it continuously moves its surface out of darkness into light and back into darkness.

Starlight without atoms can expand uniformly out of existence while atoms without starlight can remain intact.

It is the interaction of the two, an interaction that results in field replacement, that produces activity in the universe.

With the sun's immense outpouring of emissions, and the solar system's composition of planetary matter, the sun's light has a sufficient source of electrons to form inductive flows which will allow it to expand in an orderly fashion. The flows of light, while they move through empty space, do not have a lot of competition for these electrons from other sources in the environment. There are no molecules of oxygen or hydrogen, or particles of sand, or rocks which require clouds of orbiting electrons to keep them structurally sound. Atoms in darkness, on the other hand, have their clouds of orbiting electrons undisturbed by the flows of electrons that make up light and which require electrons to orbit in inductive flows in order to satisfy the excess affinity propensities of the flows.

Field replacement, the source of all activity in the universe, requires that the field of emission flows replace the affinity propensities of the nuclei of the matter that exists in the emission field.

When the atoms that make up the Earth and its atmosphere are in darkness, the affinity propensities of the nuclei that make up the atoms are in balance with the at rest motion of the electrons in their orbiting clouds.

When the flows of light are moving through space, the affinity propensities of the flows are in balance with the inductive flows that are regulating the light's expansion.

When the sunlight encounters the atoms that have been existing in peace in the darkness, it encounters atoms in which the electrons are already in use. The sunlight needs electrons to make up its inductive flows, and the only place that it can get those electrons is from the clouds of electrons orbiting the nuclei of the atoms that make up the atmosphere the sunlight is encountering as the Earth begins to turn its face toward the sunlight.

We have seen that at any point in a flow of electrons there is an excess affinity propensity capable of attracting other electrons into orbits at right angles to the flow if there are other electrons available. We have seen field replacement operate on the nucleus of an atom. The nucleus of the atom has an orbiting cloud of electrons that is satisfying the excess affinity propensity of the nucleus.

If a flow of electrons passes a nucleus, then the excess affinity propensities of the flow can replace the excess affinity propensity of the nucleus to the extent of the flow's presence, and the electrons in the orbiting cloud that had been satisfying the excess affinity propensity of the nucleus will no longer be held to the nucleus and will depart at their at rest motion, the speed of light.

The extent to which the flows, the emission field, can replace the orbiting electrons of atoms and molecules of atoms is proportional to the presence of the field. Thus, a little bit of light will replace a few orbiting electrons, while a lot of light will replace a lot of orbiting electrons. The amount of light on the surface of the sun is a lot, so much, in fact, that the field at the surface is strong enough to replace the orbiting clouds of electrons and the affinity propensities holding the units of the nuclei together, and then even the affinity propensities holding the electrons that make up the units together. On the outskirts of the Earth's atmosphere, however, there is considerably less light, although a field nonetheless.

As the incoming flows of light from the sun replace the affinity propensities of the atoms and molecules of atoms that make up the atmosphere, the nuclei begin to give off some of the electrons that make up their orbiting clouds into the ambient field that makes up the atmosphere. However, the incoming flows of light also need inductive flows, and at the margin between night and day which moves as the Earth turns under the sun, where the incoming flows replace the orbiting clouds, the electrons that made up the orbiting clouds are utilized in the inductive flows established by the incoming flows.

While the replacement is proportional to the presence of the flows of light, there are not nuclei at every point in the flow so that the process of converting over the use of electrons from orbiting clouds to inductive flows never produces a proportionate amount of electrons to satisfy the inductive flows. Thus, as the sun rises over the horizon and the flows of light burrow their way down to the surface of the Earth, their passage creates an excess affinity propensity which will attract all available electrons out of the ambient field.

If we picture any particular area of the ambient field, we will find atoms and molecules of atoms of air and any matter suspended in the air embedded in both the incoming flows of light from the sun, and the outgoing flows of low frequency radiation emanating from the surface of the Earth. The principle areas of excess affinity propensity are the nuclei of the atoms and the presence of the flows, and these excess affinity propensities are constantly in competition for the ambient electrons in the environment, electrons which are not attached to any affinity propensity, as well as in competition for the electrons attached to each other's affinity propensities. The affinity propensities of both the flows and the nuclei are stationary, although the flows can, to a certain extent, be blocked by matter as when it reflects off of the tops of the ice flecs made up of hydrogen and oxygen atoms that make up the clouds.

The electrons that make up the orbiting clouds and inductive flows have their at rest motion balanced with their affinity propensities. The electrons in the ambient field, if any, move around the area as their at rest motion moves them in response to the shifting affinity propensities in the environment. These ambient electrons are moving at the speed of light so that when an excess affinity propensity appears, their affinity propensities move them toward that excess affinity propensity at the speed of light. The creation of excess affinity propensities is felt immediately throughout the system as ambient electrons move to satisfy these areas of excess affinity propensity.

Thus, as the sunlight bores in at the margin between night and day, the excess affinity propensities created at the margin act to attract ambient electrons in the surrounding field and ambient electrons move to satisfy these newly created excess affinity propensities.

There are, however, only so many ambient electrons in the surrounding field of air, and there are just so many electrons that the nuclei of the atoms and molecules of atoms will be willing to give up to the newly created excess affinity propensities of the incoming flows so that, as the flows of light increase, the effect of the excess affinity propensities created by their presence will begin to affect the matter on which the atmosphere rests, the ground itself.

The ground, made up of solid matter, has many more atoms per area than in the gas that makes up the atmosphere and therefore has tremendous amounts of orbiting electrons available for field replacement. The same number of flows replacing the electrons in the orbiting clouds of the atoms that make up the atmosphere would be coming in contact with an infinitely greater number of atoms when it hit a sandy beach or a field of dried lava. As a result, flows that have been reduced by the distance over which they have expanded since they entered the atmosphere will replace proportionally more electrons from the surface of the Earth once they strike it.

In addition, while many of those flows will bounce off in additional expanding flows requiring additional inductive flows to satisfy their presence, depending on the crystalline nature of the surface they strike, many of those flows will continue right on into the surface and begin to dissipate within the structure itself, causing additional field replacement and additional release of electrons into the ambient field.

As the sun's flows strike the Earth, then, they replace the orbital electrons of the atoms making up the Earth. These orbital electrons are necessary to satisfy the excess affinity propensities of the flows of sunlight in establishing its inductive flows.

How do these newly released electrons get into the atmosphere to satisfy the excess affinity propensities of the flows of sunlight.

All elements and compounds of elements have a potential difference both with respect to each other and with respect to the field in which they exist. We can see this when we look at the competition for electrons in the ambient field. Elements or compounds of elements will be stable at a particular temperature because they exist in a field which is partially replacing the excess affinity propensities of the nuclei held together into the elements or compounds of elements.

The surface of the Earth exists in a field which is a combination of the field being emitted by the Earth and the emission field being emitted by the sun. Thus, stable elements and compounds of elements at the surface exist within a range of temperatures produce by night and day, with the nuclei of the atoms that make up the elements and compounds of elements having part of their excess affinity propensities replaced by the flows from the sun in the daytime, and, during the night, when the sun is no longer present, needing to attract electrons out of the ambient field to make up for the loss of the affinity propensities the presence of the sun's flows provide during the day.

We measure excess affinity propensities as potential difference, and potential differences in turn vary with the temperature, the field in which they exist. It is this potential difference between elements and compounds of elements that generate electrical flows, the movement of electrons at their at rest speed to areas where there is an excess affinity propensity with respect to the areas in which the electrons are.

Thus, the potential difference between and among the elements and compounds of elements that make up the surface of the Earth are constantly changing as the Earth moves from night to day and back again to night.

The constantly changing potential differences of the randomly distributed elements and compounds of elements on the surface of the Earth as the Earth moves from night to day under the sun is what creates the electrical flows around which atoms and molecules of atoms organize themselves into animate matter.

On a mature Earth, such as the one we occupy for the

time being, where there are sufficient atoms and molecules of atoms to organize themselves around these flows, animate matter proliferates at any point at which excess affinity propensities set up these flows on its surface. As this animate matter matures, it is subject to flows that are moving between the changing potential differences caused by the sunrise. The excess affinity propensities created in the atmosphere at the margin of sunrise as the Earth rotates affect the surface of the Earth by attracting the electrons being released from that surface as the sun strikes it, and the easiest path for those electrons to take from the Earth to the atmosphere is through the animate matter that has covered the surface of the Earth, the flowers, trees, blades of grass, morning glories.

The sunrise creates a vast charge in the atmosphere, an excess affinity propensity produced by the flows of light that require vast quantities of electrons to form inductive flows. At the same time, the sunlight replaces the need in the Earth for vast quantities of electrons, and those electrons have to find a way to move through the crystalline matter that makes up the surface of the Earth so that they can satisfy the excess affinity propensities of the atmosphere.

Empirical science measures the process as temperature differentials, but we have all seen the differential visibly in the morning, when the sunrise bores into the Earth through the cool morning air and releases more electrons from the Earth than can simultaneously migrate into the atmosphere. The result is a mirage, with the area between the ground and the horizon being turned into a mirror-like surface, most notable when the sun's rays have raised the temperature of the ground ten to fifteen degrees above the temperature of the air a yard or so above. Temperature is the movement of electrons, and when electrons in the ambient field are compressed, as they are when vast hoards of them are released simultaneously in a process which is not sufficient to produce combustion, then the effect is measurable as heat.

The simplistic model empirical science produces of the Earth's response to sunlight highlights the actual processes occurring in the Earth by being the usual one hundred and eighty degrees from reality. The current model empirical science provides of sunlight striking the Earth has the Earth taking in radiation throughout the day and giving it off again at night. If the day is filled with clouds, the Earth doesn't take in as much radiation, and if the night is filled with clouds, the night is warmer because the Earth gives off radiation which cannot radiate.

The empirical definition of radiation is rays that radiate away from a point, providing a picture of preciseness equaled only by the endless discussion of cosmic rays which clearly don't radiate, being the remnants of the big bang, and thus apparently wandering around thither and yon from this place to that. So right off the bat, empirical science has no idea what it is talking about when it claims that the Earth absorbs radiation during the day and gives it up in the night. What precisely is being absorbed, and what precisely is being given off is not a focus, especially in a scheme in which the source and effect of heat is local and its intermediate presence is merely a vibrational disturbance of a medium.

But the picture is good enough to get by, and it has permeated our recall. When light comes boring in, the Earth absorbs it, or at least absorbs the heat element of the electromagnetic spectrum, and when night falls, whatever was absorbed goes radiating back out into space once again.

Because evolution, as conceived, requires about three billion years, give or take a few days, this absorbing and radiating has to have been going on forever, so, by the good fortune that shines on the empirical mind in which everything turns out to be fact because that is the way it appears, the radiation that is absorbed by the Earth from the sun, the stars (but not the comets, which are made out of ice and thus must draw some out), and perhaps Venus exactly equals the radiation that is given off each night, right down to little Hector, who left the bathroom light on too long and had his bottom warmed but good.

This thermal equilibrium, made up to justify evolution, has little to do with reality, and nothing to do with what happens when the sun rises above the horizon.

When the sun rises above the horizon, the Earth does not take in radiation, whatever that is, it gives up electrons, and it gives them up in massive amounts!

Thus, the process that occurs when sunlight strikes matter is the opposite of what is visualized by empirical science. Nothing is absorbed, unless the increase in randomly moving electrons within the matter can be considered to be something that is absorbed. These randomly moving electrons, which we feel as heat when we touch a rock that has been sitting in the sun, are actually the net electrons left over after the light that has penetrated the surface has broken down into its individual electrons, and after the field replacement has released vast amounts of electrons from the orbiting clouds around the atoms that make up the matter, and after as many of these electrons as possible have been able to escape into the atmosphere to satisfy the excess affinity propensities of the flows of light making up the sunrise.

A hot rock, like the piece of tin, is hot because it has reached an equilibrium with the incoming light, re-emitting what it can't reflect as low frequency radiation, heat.

As the Earth rotates under the sun, allowing the land that was on the horizon at sunrise to move into the direct path of the sun's rays, the process continues. However, by the time mid-morning has been reached, the excess affinity propensities of the atmosphere above the Earth have been satisfied, meaning that, at least where there are not atmospheric disturbances such as thunderstorms, there is no longer a flow of electrons from the Earth to the atmosphere. However, the sun's light is still boring through the atmosphere and striking the ground, and doing so in a more direct path. This means that the flows become stronger toward the middle of the day.

With stronger flows comes a greater amount of field replacement, and with the greater amount of field replacement, a greater release of electrons from the orbiting clouds and a greater number of electrons left over after the light flows have broken down. When the excess affinity propensities of the nuclei of the atoms, the rocks getting hot under the sun, actually become stabilized, they begin to emit the incoming light that hasn't been reflected as light, as heat.

These electrons that make up the heat quickly become ambient, randomly moving affinity propensities in the ambient field. While there were excess affinity propensities in the morning atmosphere to absorb these electrons, as the day wears on, these electrons have nowhere to go. Ambient electrons are also heat, and as the ambient electrons build up close to the ground, where they are being emitted, so does the heat.

By mid-afternoon, there are so many ambient electrons that we begin to see mirages on the asphalt roads from the concentrations, and the excess affinity propensities next to our skin begin to draw electrons right out of the surface of our bodies, causing us to perspire.

But, as the sun begins to approach the horizon, relief is in sight. At high noon, the strongest flows of light are producing the strongest field replacement in the matter that makes up the surface of the Earth. As the surface of the Earth moves past high noon, the flows begin to weaken, lessening the amount of field replacement going on.

This allows the matter that makes up the surface of the Earth to start to reclaim the electrons out of the ambient field to replace the excess affinity propensities of the nuclei of the atoms as the sun's emissions recede.

The atoms that make up the atmosphere are also beginning to obtain excess affinity propensities, but in this case, the ratio is reversed from the sunrise. During the sunrise, the flows required proportionally more electrons to make up the inductive flows than could be given up by the atoms whose affinity propensities they replaced because of the spatial difference between the area covered by the flows and the lesser area affected by the nuclei of the atoms. When the sun begins to depart the atmosphere, however, the release into the ambient field of the electrons that made up the inductive flows grossly exceeds the number the atoms need to satisfy the excess affinity propensities resulting from the departure of the flows. This differential was what made the morning glories open up and the morning mirages blossom and it now causes a build-up of electrons in the ambient field that can be seen as heat lightning in the dusky distance.

In the meantime, with the departure of the sun and its flows of light, field replacement on the surface of the Earth has come to an end, and the Earth is put in a position of having as much excess affinity propensity needing orbiting electrons as the morning atmosphere had.

It is not as easy, however, for the electrons to find a path back into the Earth as it was for them to find paths out of the Earth. Morning glories are going to close up before they become lightning rods for the return of ambient electrons. Afternoon and evening thunderstorms might allow lightning to discharge large amounts of electrons into the ground, but the basic process is just the slow reclamation of the electrons as the night goes on, with the temperature dropping off as the excess affinity propensities of the Earth slowly drain the ambient electrons from the atmosphere.

The less material areas of the surface accept the influxes the most readily. Thus, bodies of water, porous to the onslaught of electrons, find themselves the recipient of such vast quantities of electrons that the electrons create flows that actually replace the affinity propensities of the nuclei of the water molecules, breaking them down and causing them to form steaming mists on the surface.

The process, removing ambient electrons from the atmosphere, is reducing the number of those ambient electrons and we have the opposite of heat, cooling.

Cooling is actually the result of matter losing electrons. If our skin has less of an excess affinity propensity than the surrounding air, the excess affinity propensities of the air will draw the electrons right out of our skin, just as an excess of electrons did. The similarity of effect provides the basis for an old trick, blindfolding someone to see if he can tell the difference between a burn and a piece of ice. Presumably having remembered the experience of the burn, the person will feel a piece of ice just like a hot iron. The effect of both is the same. The hot iron replaces the electrons in the flesh, causing them to depart. The ice pulls them out physically because of the massive deficit of electrons it has with respect to the skin. This is why meat gets freezer burn and frost bite actually burns flesh off.

Of course, the night can become colder because cold air masses, with their electrons already removed, move into the area as a result of the Earth's rotation, and these cold air masses will have the same effect on us as being burned by the ice. They'll pull massive amounts of electrons out of our body and cause us to freeze to death. But the warmth that continues into the wee hours on a cloudy night does not result from the Earth's radiation being blocked from traveling into space thereby allowing the surface to maintain a warm temperature. The opposite is happening. The flow of electrons is in the other direction, from the atmosphere to the Earth.

What keeps things warm on a cloudy night is the electrons in the clouds themselves. As the Earth draws back the electrons from the ambient field, any excess affinity propensities created at the surface of the Earth, affinity propensities that would be measurable as cooling, pull electrons out of the atoms and molecules of atoms that make up the clouds. Thus, the clouds provide a buffer for the removal of electrons from the atmosphere close to the Earth, replacing those ambient electrons so that the cooling doesn't occur. Under the right conditions, the removal of electrons will cause precipitation and we'll get the dew that drapes the night in preparation for another onslaught of the sum's rays.

When the sun comes up, the Earth doesn't absorb

radiation, it emits electrons, making the day hot. When the sun goes down, the Earth doesn't emit radiation, it absorbs electrons, making the night cold.

If the night is cloudy, the electrons removed from the night air are replaced by electrons from the clouds, allowing the night to remain warm.

Where do the clouds get their store of electrons?

Clouds made up of hydrogen and oxygen atoms, separated from water molecules at the equator in a process that causes the separated nuclei to attract large orbital clouds of electrons, are a vast storehouse of energy, the orbital clouds, which are emitted in field replacement when the hydrogen and oxygen atoms recombine into water molecules toward the poles in what is one of the most energetic chemical combinations in matter.

12. Light In The Ambient Field

A thunderstorm is a thing of intellectual elegance.

Weather is produced by the temperature differential at the equator and the poles. The nature of the relationship between the hydrogen and oxygen atoms, their ability to form water and ice, and their resulting ability to take on vast numbers of orbital electrons when they separate at the equator, and give them off when they later recombine into water molecules, allows these atoms to transport the equatorial energy north and south toward the planetary poles where the atoms recombine into water molecules and precipitate out as rain releasing vast clouds of electrons in the process.

Water is broken down at the equator into hydrogen and oxygen atoms which rise in the equatorial air masses, leaving room for the arrival of warm, electron laden air from the poles. This air, warmed during its southward movement, provides, along with the direct equatorial sunlight, the source of electrons needed to fill in the orbiting clouds of the individual oxygen and hydrogen atoms as they split apart from the water molecule.

The oxygen and hydrogen atoms, ice flecs by the time they gain several miles in altitude, reach their own level of displacement in the atmosphere and are pushed north by the continual process of molecular splitting occurring at the equator. These ice flecs are bundles of frozen hydrogen and oxygen atoms which have risen into geometrically increasing areas. As the atoms rise from the equator, their affinity propensities, moving into greater areas which contain less affinity propensities, draw the individual atoms together at the same time that the individual nuclei draw their massive clouds of orbiting electrons into tighter orbits, creating stable atoms that move in the upper atmosphere. When these atoms are forced back down into the increasing field of the lower atmosphere, the atoms begin to recombine back into water molecules, releasing the massive number of electrons as heat.

What exists normally as a water molecule on the surface of the Earth, has broken down into individual atoms. These individual atoms have obtained massive numbers of orbiting electrons which move close to the nuclei as the atoms rise into the atmosphere.

As electrons are heat, the ice flecs are transporting the equatorial heat to less temperate regions. As the ice flecs contain oxygen and hydrogen atoms, they are capable of reforming into water molecules.

The planet is rotating on its axis at a uniform rate. The circumference of the Earth becomes smaller towards the poles. As a result, a point on any circumference parallel to the equator takes the same time to move a shorter distance the closer the circumference is to the poles, and thus moves slower until the circumference reaches the pole at which point it becomes a point, and doesn't move at all.

When the masses of hydrogen and oxygen flecs begin their journey, in our hemisphere, north, they are moving at the speed imparted to them at the equator, about a thousand miles per hour. However, the speed of the surface of the Earth diminishes as the masses travel north.

As a result, air masses starting out at the equator traveling a thousand miles per hour, with nothing but a decreasing volume to retard their motion, outpace the surface of the Earth, giving them a westeast motion relative to the ground. As the air masses are also moving north, the overall motion with respect to the ground is northeast. As the area available for their movement becomes geometrically less the further north the air masses move, the air masses are forced down into the lower atmosphere. When they are forced down into the lower atmosphere, the field in which the oxygen and hydrogen ice flecs that make up the air masses exist increases, and the ice flecs begin to come together into water molecules once again, forming clouds and releasing the vast stores of electrons they obtained at the equator when they split apart.

Weather is not a phenomena subject to mathematical analysis, although it can be fairly repetitive for specific locales. The production of thunderstorms is no exception although specific conditions always produce the perfect thunderstorm.

There has to be a sunny day which has produced flows of light that have warmed vast stretches of ground. As a result of field replacement, the ground has given up vast numbers of electrons into the atmosphere. As long as the sun is shining, the incoming flows of sunlight balance the excess affinity propensities of the ground with nonreflecting flows of light being emitted as low frequency radiation, heat. If these flows of light were somehow disturbed, cut off, then the excess affinity propensities of the ground would re-emerge with a vengeance, demanding a return of all of the electrons that were being replaced by the sun's rays.

Without any weather, the day would proceed, the Earth would rotate past high noon, the flows would diminish, the vast areas would slowly regain their deficit of electrons to satisfy the rising excess affinity propensities, and the cooling breezes of night would arrive.

However, if there are masses of hydrogen and oxygen flecs moving northeast, and these ice flecs are forced down into the warmer atmosphere, the conversion of the ice flecs into water molecules begins, and all of a sudden, clouds appear out of nowhere.

The conversion process involves massive numbers of hydrogen and oxygen atoms in the form of ice flecs being forced down into the lower atmosphere where it is warm. The forming molecules of water are pushed aloft in vast thunderheads by the force of the downward moving ice flec laden air. As the momentum of the ice flecs is toward the northeast with respect to the ground, the forming thunderheads also move toward the northeast with respect to the ground.

While the conversion process goes on in the warmer atmosphere closer to the Earth, and the warmer atmosphere is warm because of a high number of electrons existing in the ambient field, the conversion process releases vast stores of electrons into the same field. As a result, the number of ambient electrons in the area around the forming clouds becomes saturated beyond their physical capacity.

The forming clouds, however, are having a far more dramatic effect on the ground beneath them. They are blocking the flows of sunlight from reaching the ground. As soon as the sunlight is blocked from reaching the ground, the flows of sunlight cease to replace the affinity propensities of the nuclei of the atoms that make up the ground, and these nuclei are all of a sudden put in a position of having excess affinity propensities that need electrons to satisfy them.

Where are the electrons?

Vast numbers are being released in the areas around the forming thunderhead as the ice flecs of oxygen and hydrogen combine into water molecules. This is the opposite of what is termed evaporation, which eliminates heat, and is called condensation, which produces heat. Evaporation of water molecules at the equator and the condensation of atoms of hydrogen and oxygen from ice flecs in clouds are not molecular processes, the breaking down of water molecules into smaller water molecules, the weather analogy of molecular magnets. They are the processes involved in the breaking down of water molecules into individual atoms of hydrogen and oxygen and recombining atoms of oxygen and hydrogen into water molecules.

As a result of the recombination of the hydrogen and oxygen atoms, a process that eliminates the need for many of the orbiting electrons that had existed in the orbital clouds of the individual atoms, there is an intolerable number of electrons around the clouds.

With the sunlight blocked from reaching the ground, there is an incredible build-up of excess affinity propensity in the ground.

Ambient electrons move to satisfy excess affinity propensities, and it only takes a sufficient differential to overcome the resistance provided by the air and the crystalline structure of the ground that separates the excess affinity propensities of the ground from the massive number of ambient electrons surrounding the clouds.

The ambient electrons organize themselves against the resistance, and find a path to the ground, resulting in lightning.

Thus, the cloud, converting hydrogen and oxygen atoms into water molecules, gives off unneeded electrons from the orbiting clouds of the atoms as their nuclei move closer together to satisfy each others excess affinity propensities while blocking the sunlight from continuing to replace the excess affinity propensities of the nuclei making up the ground beneath. The unneeded electrons find their way to the ground to satisfy these newly created excess affinity propensities in the form of lightning.

As the storm moves northeast, the clouds move out of the way of the sun's rays, and, at the tail end of the storm, the rays once again flow into the Earth, replacing the excess affinity propensities of the atoms that make it up, and everything returns to normal.

Unless there is a feedback of the newly released electrons at the trailing edge of the cloud!

The thunderstorm is a giant engine that marches across the land, converting oxygen and hydrogen atoms moving northeast from the equator into water molecules which wash the fruited plains. The northeast moving air masses containing these atoms power the mechanism as they are forced into the lower atmosphere, and their conversion produces the electrons which are dissipated as lightning into a ground made electron deficit by the blocking effect of the cloud on the sun's rays. When the cloud moves on, the sunlight replaces all of the orbital electrons obtained from the lightning and the ground has the same relation to the atmosphere as it had at sunrise, an excess number of electrons.

If the potential difference between the ground and the atmosphere in the wake of the storm is sufficient, if there is enough of an excess of electrons trailing the thunderstorm in the ground and a great enough excess affinity propensity left in the atmosphere in the storm's wake, the electrons in the ground will begin to mimic the sunrise and move massively back into the air.

To see the result, we have to go back and remember what is always associated with a flow of electrons. Light is a structured flow, and electricity is a massive, nonfrequency flow. It is the latter, massive nonfrequency flow that occurs in the wake of a thunderstorm.

Inductive flows are associated with flows of electrons and inductive flows move at right angles to the primary flow. Thus, as the electrons return to the atmosphere in the wake of a thunderstorm, they establish inductive flows. We can use the right-hand rule to determine the direction of these inductive flows. Pointing the thumb of the right hand up, the fingers curl in a counterclockwise direction. These inductive flows, if sufficient, capture the atoms and molecules of atoms that make up the air, just like a glass rod captures bits and pieces of paper, and the resulting force causes the air to start to move in the same direction.

If sufficient atoms and molecules of atoms are caught in the vortex of these inductive flows, they will begin to spin around the returning electrons, producing the counterclockwise funnel we see as a tornado. The atoms and molecules of atoms available for creating the vortex are the newly created water molecules at the tail end of the storm. Because the tail end of the storm has clouds that still block the sun's rays, the electrons released by the newly revealed sunlight behind the storm race under the tail and then leap into its underside, giving rise to the funnels that can become tornadoes.

These tornadoes are machines in their own right, returning the massive numbers of no-longer-needed electrons injected into the ground by the lightning of the thunderstorm back into the atmosphere as they trail under the tail end the thunderstorm as the thunderstorm moves northeast.

Occasionally, the reverse process occurs in the storm itself if flows of electrons can be established while the clouds are converting ice flecs into water molecules while blocking the sun's rays from striking the ground. Instead of lightning discharging the excess electrons from the cloud to the ground, the storm might establish a massive flow of electrons. Once again, using the right-hand rule, pointing the thumb from the cloud to the ground would give us the direction of the resulting tornadoes formed, clockwise funnels. The reason most tornadoes rotate counterclockwise is because the sunlight at the rear of the storm replaces vast numbers of electrons which can readily organize themselves into flows whereas the movement of the cloud tends to discourage attempts to establish a solid flow to the ground.

This magnificent picture of the interaction of light in the ambient field, the thunderstorm, gives us an idea of the competition among excess affinity propensities in which we live and of which we are a part. Light is a balanced structure with the affinity propensities of the expanding flows of electrons balanced by the affinity propensities of electrons orbiting in inductive flows around the primary flows. In empty space, light will expand undisturbed until it diminishes out of existence, dissipates into its individual electrons moving randomly as their at rest motion chases available affinity propensities.

Matter is made up of atoms which are balanced structures composed of the same electrons, with all affinity propensities used either to hold the nuclei that make up the structure together, or to hold electrons in orbit around the structure. If the atom existed in empty space, devoid of flows of light, it would stay in balance forever.

When the two interact, light with matter, matter with light, there are shifting affinity propensities because stationary affinity propensities replace moving affinity propensities, and some affinity propensities are more stationary than others. Shifting affinity propensities produce heat or energy and remove heat and energy from the environment. Just by observing a process as simple as a burning match, we can see that light replaces affinity propensities which in turn produces light, replaced affinity propensities.

Because we exist on the surface of a planet that is rotating in front of an emitting body of matter in such a way that most points on its surface pass in and out of the sunlight on a regular basis, and because the rotation creates an equatorial plane that is roughly in line with the equatorial plane of the sun, the temperatures vary from pole to equator, and because the planet's own combustion process has receded beneath the crust to a point that it has very little affect on surface temperature, making the major player affecting that temperature equatorial rotation, we exist in a changing field which is continually altering the excess affinity propensities of the matter within the field.

As a result, flows of electrons are constantly moving back and forth between and among those shifting excess affinity propensities.

In any given space there is a hierarchy of affinity propensities. There are the nuclei of atoms of matter made up of units which are balls of electrons held immobile by their affinity propensities, all of their at rest motion potential. These units are at the top of the heap when measured by resistance to being torn asunder. They exist unless a field is so strong that it can replace the very affinity propensities holding their electrons together surface to surface, the strongest affinity propensity possible, an affinity propensity with no at rest motion affecting it.

Next in the hierarchy comes the nuclei themselves, in which units are held together by their excess affinity propensities. The field replacement required to allow the units in a nucleus to separate are less than needed to unravel the electrons from a unit, but still beyond the reach of normal activities.

The nuclei themselves are held together by their own excess affinity propensities, nuclei to nuclei, as distanced by the clouds of orbiting electrons that satisfy any excess affinity propensities left over after the nuclei come together as closely as possible. It is these affinity propensities that we manipulate by stripping atoms of their orbiting clouds, and from each other, mixing them with other atoms similarly stripped, and recombining the results in chemical reactions. We are, with this technology, manipulating affinity propensities to produce something that didn't exist when the matter that makes up the crust broke down in the planet's normal process of cooling.

We have any number of ways to replace the affinity propensities.

We can compress atoms. Compression forces the nuclei closer together mechanically, causing them to replace each others excess affinity propensities. The orbiting electrons they give up from their clouds are measured as heat, heat that results from compressing a gas. We can place atoms in solution and heat the solution. We can create flows of electrons with generators which, when applied to matter, cause field replacement to occur. We can concentrate electrons by bouncing them back and forth to produce lasers which can be directed without conductors to produce field replacement in matter.

All of these processes take technology to accomplish, and involve producing, on a local basis, excess affinity propensities that far exceed the tolerances that normal matter occupies, the tolerances being the lower and upper ranges of the field in which the matter can exist without being substantially affected by the field. The animate matter that makes us up does fairly well between the temperatures of fifty and ninety degrees, ice does well under thirty-two degrees Fahrenheit, water exists between thirty-two degrees and two hundred and twelve degrees Fahrenheit, lead loses its shape at about 620 degrees Fahrenheit.

Light itself, though, does affect the state of the atoms that make up the atmosphere and it does affect the state of matter on the surface of the Earth. And while we have seen that shifting affinity propensities can have some pretty spectacular results, there are a lot of things going on around us that exist without our particular awareness. We broadcast invisible frequencies from points all over the landscape, each of these broadcasts being expanding flows of emissions with frequencies established by the cycle of their creation. We string power lines from one place to another, each containing flows of electrons which alternate, expanding and collapsing the inductive flows around them in the process while being subject to drawdown by users, which means that electrons are constantly falling out of the flows and having to be replaced. We have step up and step down transformers located around users which alternately take in and give up electrons, and the users themselves turn on lights that bounce structured flows of electrons off walls, activate burners which spew them out when they can't find their way through nonconductors as heat, and switch on flows that pepper phosphorous tubes into constantly emitting more electrons.

The electric light is the most obvious example of electrons being converted into structured light.

Electrons can be made to move in a conductor by making the conductor a path between two differences in potential so that electrons from the area with less of an excess affinity propensity flow at their at rest speed to areas with a greater excess affinity propensity. Electrons can also be made to move in a conductor by tipping them into the conductor from secondary flows of electrons, either the inductive flows of other conductors, or the external flows around magnetic material.

Tipping electrons into a circuit from magnetic lines of force or from the inductive flows of a conductor depletes the electrons in those magnetic lines of force or inductive flows. This results in a deficit of electrons in the magnetic lines of force or inductive flows, an excess affinity propensity that results in the magnetic lines of force or inductive flows drawing electrons out of the ambient field.

For every electron tipped into the conductor, an electron has to come out of the ambient field!

A generator is simply a medium through which electrons in the ambient field are converted to electrical flows in conductors.

Thus, the technologies we create to move electrons from one place to another are based on the creation of excess affinity propensities that continually draw electrons out of the ambient field to replace the electrons tipped into the conductor.

If we buy a toy car that is battery powered, placing a battery in the car gives the motor in the car a source of electrons. The electrons are stored in the battery by placing two cells of different matter insulated one from the other, each of which has its affinity propensity in balance with respect to the field in which it exists, but each having a potential difference, a relative difference in affinity propensity with respect to each other. As a result of this potential difference, when the two cells are connected, electrons flow through the coils of the motor, creating inductive flows which push the coils away from the external flows around magnets to produce rotary motion in the motor which in turn turns the wheels of the car.

Few of the electrons are lost into the ambient field because it is their presence, as they move through the conductor, that creates the inductive flows that push against the existing flows. The electrons that make up the inductive flows are drawn from the ambient field and are not being used for any other purposes.

Lowering the field in which the system is operating will reduce the electrons available in the ambient field to form the necessary inductive flows. The motor may not work in frigid temperatures, but it is not because the battery doesn't still produce a potential difference that can establish an electric flow. It is because the motor can't draw electrons out of the ambient field to produce inductive flows which can push it in a rotary manner.

Lowering the field in which the battery exists, on the other hand, also alters the relationship of the different elements or compounds of elements that make up the cells with the field, and with respect to one another so that at very low temperatures there might be no difference in excess affinity propensities, and nothing to attract electrons. On a very cold day, the battery won't put out juice for the motor because the lowered field has eliminated the potential difference that provides the basis for that juice.

How the technology we use to manipulate potential differences is dependent on the existence of electrons in the ambient field and on the field itself.

If we take the motor out of the toy car, we can spin its rotor. Spinning the rotor has the effect of moving conductors, the windings of the rotor, through the lines of force produced by the motor's magnets or the inductive flows if the motor uses electromagnets. This causes the electrons from the external flows to be deflected into the conductor that makes up the motor's coils.

For every electron deflected out of the external flows into the conductor, the magnetic lines of force have to attract another electron from the ambient field.

If we are just moving a wire through the external flows, the deflected electrons do little but move through the conductor and back into the ambient field, replacing the electrons the magnetic lines of force attracted out of the ambient field to replace the ones lost to the conductor. If we move a conductor that is closed in a circuit through the external flows, we will defect electrons into the conductor, and they will move around the conductor in a circle at their at rest motion. However, as soon as all of the electrons that can move around in the conductor are moving around in the conductor, no more electrons will be deflected into the conductor. Stopping the movement of the conductor will allow the electrons tipped into the conductor to immediately escape the conductor and rejoin the ambient field.

Placing a load on the conductor, something like a light bulb that can use up the electrons, will result in the measurable flow of electrons that we call electricity.

If we have a load on the conductor, if we connect our little car motor up to a light bulb, then spinning the rotor will cause the conductor to pass through the external flows of electrons around the magnets in the motor which will result in those electrons being tipped into the conductor. Once in the conductor, the electrons will move around in the circuit provided by closing the conductor with the load. The load will remove the electrons tipped into the conductor from the magnetic lines of force.

As the load removes electrons from the conductor, the conductor's movement through the motor's magnetic lines of force tips electrons out of those flows into the circuit to replace the electrons lost to the load. The motor's magnetic lines of force then have to obtain replacement electrons out of the ambient field, and, providing there are sufficient ambient electrons in the ambient field, they will become a part of the motor's magnetic lines of force.

The motor, operating as a generator, is actually converting electrons from the ambient field into electron flows by tipping electrons out of the magnetic lines of force established by its magnets, either magnetized matter or electromagnets, into conductors and then making up the deficit of electrons in its magnetic lines of force by attracting electrons out of the ambient field. Passing a conductor in a circuit with a load through magnetic lines of force removes ambient electrons from the area around the conductor and puts them to work in the load.

If the motor is really operating efficiently, has a huge load and a superman spinning its rotor, the rate at which electrons are drawn out of the magnetic lines of force, and the rate at which they are replaced might well deplete the electrons available in the ambient field. When this occurs, the excess affinity propensities attracting electrons out of the ambient field might become greater than the affinity propensities holding orbiting clouds of electrons to the atoms and molecules of atoms in the air surrounding the motor.

At this point, the excess affinity propensities of the motor, operating through its magnetic lines of force, will begin to strip electrons from the orbiting clouds of the atoms and molecules of atoms around the motor, leaving the nuclei in the surrounding air with excess affinity propensities.

Nuclei with excess affinity propensities are measured as ionized atoms, atoms measured with a charge, not the classical positive or negative charges, which in any event, have no qualitative difference, but simply the excess affinity propensity that results from having a deficit of electrons.

What is actually happening is that the hierarchy of affinity propensities is causing electrons to move to where there is the greatest excess of affinity propensity, and when the operation of the motor draws more electrons out of the surrounding air than are available without affecting other existing affinity propensities in the area, then the operation of the motor will begin to pull electrons out of the orbiting clouds of the atoms and molecules of atoms that make up the air because it is producing excess affinity propensities greater than those holding the electrons to the nuclei of the atoms and molecules of atoms making up the air.

The nature of the load determines the rate at which electrons are pulled from the conductor. Operating an electric motor results in only minimal loss, usually through the heat involved in pushing the rotor against the friction of the motor's parts, although when the motor is in turn pushing something, the loss of electrons is proportional to the amount of resistance, work. The opposition of the flows of electrons doing the pushing might become so great that they begin to leave the flows en masse, producing heat that can organize itself into combustion, burning up the motor.

An incandescent light bulb provides an example of how a load draws electrons out of a circuit and how light is created at a single frequency. The bulb is made up of a filament in a vacuum. The filament is made up of atoms whose affinity propensities are so strong that they can retain their crystalline structure when all of the electrons in their orbiting clouds have been replaced.

When not emitting light, the filament has its nuclei formed into matter with orbiting clouds of electrons balanced by the attraction among the affinity propensities of the nuclei.

However, when a current flow is passed through the filament, all of the electrons in the orbiting clouds are replaced and emitted, leaving behind excess affinity propensities in the nuclei of the filament. These excess affinity propensities immediately begin to remove electrons from the flow of electrons passing through the conductor, but none can take up positions in the orbiting clouds without immediately being replaced by the electric flow itself.

The nuclei continually attract but fail to fully capture orbiting electrons out of the flows of electricity. The electrons are emitted by the filament before they can become orbital clouds. As long as they are emitted, more electron are drawn from the current flow. As electrons build up on the surface of the filament, the nuclei that make up the filament move apart, and the electrons are emitted, causing the nuclei to move back together. Additional electrons are drawn out of the flow of electricity, and as they build up in the filament, the molecules that make up the filament move apart and the electrons are emitted. We saw this process in both the combusting conductor and the combusting head of a match.

The process creates a frequency established by the strength of the flow of electricity and the nature of the atoms making up the filament, and it is this frequency that is emitted as the white or yellow light, a single frequency dictated by the rate of combustion. The reason that the process has to be carried out in a vacuum is that, just as in the example of the combusting conductor, atoms next to the filament would become ionized, obtain excess affinity propensities from the process, and seek to combine with the atoms in the filament, causing the nuclei of the atoms in the filament to come apart, breaking the cycle of emission and burning out the light bulb.

Alternating current generators can generate electricity for transmission over greater areas than direct current generators because they require less of a potential to remove electrons from the ambient field around the generator at one time. The only difference between a direct current generator and an alternating current generator is the direction electrons are tipped into the conductor. With a direct current generator, the electrons pushed into a conductor have to match the probable loads whether the loads are there or not, and thus the potential size of the system is limited by the generator's need to be ready to do useless work, tip electrons into the conductor that currently have no place to go.

Alternating current generators tip electrons into the conductor in one direction, and then the other direction, in regular cycles. This allows electrons in the conductor to move continuously back and forth in the conductor so that it makes no difference whether potential loads are being used or not, and the number of loads therefore does not limit the size of the resulting system.

Placing a load on an alternating current circuit

removes electrons which are immediately replaced by the generator. Drawing down on the overall circuit produces the same result that putting a load on an electric motor produces, the electrons drawn out have to be replaced.

Where do the electrons come from that are tipped into generators located in the vast concrete rooms of dams or on the floors of fossil fuel or hot rod generating plants?

The fact that these generators, weighing multiples of tons, have to be bolted to the floor or sunk in concrete to operate provides a clue as to the source of these electrons. While the movement of vast quantities of water over a dam produces a ready source of electrons, as does the combustion of fossil fuel, and the cooling ponds of nuclear plants, the greatest single source of electrons for generating plants is from the Earth's emission field itself.

The heat output at the surface of the Earth is a result of its cooling process, and is estimated to be in the area of six trillion calories per second. This heat is made up of low frequency radiation, flows of electrons with their associated inductive flows. This emission field provides more than enough electrons to replace those leached out of the ambient field by a generator's operation.

The operation of generators removes electrons from the ambient field, and the operation of huge generators removes electrons from the inductive flows that recombine the expanding flows of the Earth's emission field to produce the mechanism that regulates the uniform expansion of that emission field.

Removing the electrons available to make up the inductive flows that produce the mechanism that regulates the uniform expansion of the Earth's emission field eliminates the mechanism itself.

The mechanism that regulates the uniform expansion of the Earth's emission field has a by-product.

That by-product is the force that moves the units in the nuclei of the atoms within that field back toward the source of the emission field, the surface of the Earth.

Without the mechanism that regulates the uniform expansion of the emission field, the by-product of that mechanism, the force that moves the nuclei of atoms back toward the source of emissions, also becomes irregular.

Because generators leach electrons out of the inductive flows of the Earth's expanding emission field, their operation creates irregular pockets of gravity around the generator. These irregular pockets of gravity lighten the matter than makes up the generators, and makes their operation erratic as a result.

While most of the bucking produced by a generator operating at full tilt is laid down to the mysterious processes of torque produced by the spinning shaft of the generator, bolting the generator down or sinking it in concrete masks the fact that the effect of the torque on the generator producing electricity is far greater, but more important, far more irregular, than the effect of torque on a nongenerating generator.

Like good monkeys, we monkey saw the generators buck more the bigger they got, and we monkey said torque. As soon as we had the effect categorized as torque, we had an effect whose cause we could regulate. Bouncing matter could be made to stop bouncing by bolting it to or encasing it in something that didn't bounce.

It didn't really make any difference what the cause actually might be, because we are only dealing with categorizing effects.

In like manner, it doesn't matter what the cause of light is, it doesn't matter what the cause of electricity, magnetism or heat is, it doesn't matter what the cause of planetary orbiting or rotation is, or what actually causes objects to fall. As long as we can monkey see the sun and monkey say light, as long as we can monkey see falling objects and monkeys say gravity, we have fulfilled our goal as monkeys.

We can sit around chattering among ourselves what to name various effects and what to include in the name and we will be happy little monkeys.

That is, after all, the goal of empirical science, to answer our questions so that we never again have to ask them and can thus die happy.

Isn't it?

13. To See Or Not To See

Light conveys the information about the environment that we reform as pictures with our minds. In accordance with the dictates of empirical science, which satisfies itself with classifying things, giving things names, and then mistaking the names for comprehension, we can make the statement that a light source illuminates objects allowing us to see those objects.

The statement says absolutely nothing about how this happens, the hallmark of empirical statements about physical reality.

Physical reality is made up of points that are the boundaries of crystalline surfaces. Light from a source is actually an expanding flow of light. The expanding flow of light diminishes uniformly over the surface of an expanding sphere whose center is the source of the light. When the light strikes points in physical reality, it bounces off of those points. Each point off of which light bounces produces another expanding sphere of light which also diminishes uniformly over the surface of an expanding sphere whose center is the point off of which the light has bounced. The light also has a frequency which is reduced in accordance with the nature of the surface of the matter off of which it has bounced.

Just as we can sit with a crystal and tune in all of the expanding spheres of radio transmissions that exist in the atmosphere around us by their frequency, so too do our eyes register the frequency of the expanding spheres of light bouncing off of the points that make up the objects around us.

More important, our eyes register the relative distance to those points from the surface of our eyes because the expanding flows of light, by their intensity, contain information about the relative distance of all the points off of which light is bouncing in physical reality.

There is nothing mystical about this process. It is simply a mechanical process that results from animate matter organizing itself to take advantage of the physical conditions in the environment. Those physical conditions consist of flows of light that bounce off of the hard edges of physical reality, and in the process, convey information about those hard edges to organisms that have evolved to take advantage of the information.

The information is the result of the simple geometrical nature of light's movement in physical reality, and the manner in which the light was produced, with a frequency of emission that is alterable by its interaction with the matter off of which it bounces and an intensity that diminishes uniformly with distance. The information about distance allows us to locate ourselves with respect to objects in physical reality, and thus to move successfully within that physical reality, and the frequency changes allow us to distinguish surface features within that physical reality.

Seeing physical reality is one thing, knowing about it is an entirely different thing.

We can craft chains and rods, scales and sundials to measure those things that the eye can see.

Understanding the nature of what we see, however, requires the mind.

The eye is merely a detector that evolved to locate the organism to which it is attached in physical reality. The only way that an organism can locate itself in physical reality is by field replacement. The most elemental mechanism that operates by field replacement controls the cilia that paddle simple organisms away from heat. As an organism drifts into warmer water, field replacement releases electrons from cells which activate the paddles and the organism reflexively moves back into cooler waters. Our own skin is covered with similar mechanisms. If we approach something that is too hot, field replacement releases electrons which activate our muscles to move us away from the heat without the mind mediating the transaction.

The field replacement that occurs in the eye is somewhat more complex, but then the eye has to form units containing information about all of the points off of which light is bouncing in physical reality, units which then move up the optic nerve to the organic matter of the brain where the information is decoded and emitted as electron flows through the equilibrical structure of the mind.

In discussing light, we need to know that the eye works as a detector, a platform attached to an organism which is able to move among the various flows that are bouncing off of physical reality and, because the flows contain information about the relative distance of those points from the platform of the eye, is capable of locating the organism with respect to all of the points in physical reality.

The pictures of physical reality encoded in the eye and decoded by the brain are passed through the equilibrical structure of the mind. The pictures emerge from the effect of the electrical flows on the dynamic structure of the mind which is made up of electrons held into a web of stable equilibrium by their opposing properties of at rest motion and affinity propensity. The electrical flows, incapable of causing the electrons that make up the structure of the mind to form inductive flows, merely influence the electrons to move from their positions of stable equilibrium in proportion to the strength of the electrical flows passing through the structure. Because the movement of the electrons that make up the equilibrical structure of the mind is proportional to the flows moving through it and because the flows represent the relative distances of the hard edges of reality from the eye, the mind can form fairly objective representations of reality.

The electrons that make up the structure of the mind, however, do attempt to return to their positions of stable equilibrium, and in doing so, they produce a second flow of electrons that approximates the flows that disbalanced them. This second set of flows is recaptured by the organic matter of the brain and re-encoded in a unit that is stored in the neurons that make up the majority of the brain's organic material.

The process of decoding and re-encoding the information from the senses produces a current in the brain that is equal to the flows which are forming pictures in the mind as they move through the mind. As these electrical flows have presence, they have excess affinity propensities.

The excess affinity propensities of the flows formed in the eye when physical reality is encoded equal the excess affinity propensities of the encoded units that represent the flows. The excess affinity propensities of individual units thus match the excess affinity propensities of the flows.

The excess affinity propensities of the flows carry the encoded unit into the brain for decoding. The re-encoded unit has approximately the same excess affinity propensity as the original encoded unit, so that when the current flowing throughout the organic matter of the brain is carrying encoded units from the eye for decoding in the brain, and re-encoded units from the brain into the neurons for storage, it is also picking up any re-encoded units in the neurons that have similar excess affinity propensities.

These units, picked up from the neuronic storage cells, constitute recall, and are mixed in with the encoded units coming from the eye for decoding by the brain.

It is only when the re-encoded units coming from the neurons, recall, and the encoded units coming from the eye, reality, agree with one another and thus can be reformed in the mind simultaneously that we can comprehend the physical reality that we see, that our recall matches what we see in reality and thus provides us with information about what we see.

We recall what we see and not something else because what we see generates a current level at which our recall is stored. If we have nothing stored in our neurons, if we have no recall for what our eyes are seeing, we have to stop and try and figure out what we are seeing, because without recall, we can't see what is right in front of our eyes.

To see the light in reality, we have to see the light in recall.

If we have nothing in recall that allows us to comprehend what we see when we see light, then we will not have any recognition of what the light is.

If we have something in recall that allows us to comprehend the light, but that recall has nothing to do with what the light actually is in physical reality, then we will be able to see only what reflects the light, and not the light itself.

Details about how the mechanical mind forms, stores and recalls pictures of physical reality can be found in the last four chapters of *Atoms, Stars and Minds,* while how the mind generates behavior when recall and reality disagree can be found in *The Model Mind,* the fourth volumes of The Copernican Series. We are here interested in how recall operates in conjunction with the information the eye receives and transmits to the mind to produce comprehension, the light bulb that goes on in our minds when we connect something in reality with our recall so that we literally see the light, we understand. All the light in the world will not illuminate something if we don't have the information in our recall with which to compare reality so that we can deal effectively with that reality.

When the big mistake was made, when it was found that our minds had deceived us into believing that the sun traveled around the Earth when it was actually the Earth that was traveling around the sun, we told ourselves that if we limited our observations to measurements, then we could never go wrong, we could never be tricked by our minds again.

However, measurements in and of themselves have no meaning. Seeing white light enter one side of a prism and exit the other in different colors doesn't tell us anything about what is going on when the light interacts with the matter that makes up the prism. We don't know what light is, we don't know what the stuff that makes up the prism is, we don't know what happens when the light interacts with the stuff that makes up the prism, and we don't know what is coming out, what the significance of the individual colors are.

No matter what we do, we will never be able to come up with tools that will allow us to measure directly what is happening with light. This isn't an unfamiliar situation. We have no way of fabricating rulers that will allow us to directly measure the Earth moving around the sun. This does not stop us from attempting to measure because when we see something that we can't directly measure, we are still driven to comprehend it.

Because we are driven to comprehend, to understand, and because comprehension is merely the agreement of our recall with what we measure to exist in physical reality, we can make up anything we want to explain physical reality so long as what we make up doesn't disagree with the measurements we have made of physical reality. Thus, we create the monkey see, monkey say process of empirical science in which all we have to do is describe what we see.

The major mistake of empirical science is believing that measurements provide comprehension. We see something in reality, we have nothing in recall with which to compare it, and we look around in reality for something else with which to compare it. This process of classification, at the basis of the empirical process, is merely the process of placing in our neuronic storage bins recall that will match what we see in reality.

We are storing things at similar current levels.

After we have something in our recall, our recall will match reality and when we recall what we see, we believe we have comprehended. Having something in recall that matches what we see in reality provides the feeling of knowing.

Classification, creating recall with similar excess affinity propensities recallable at similar current levels, produces categories in which the things we see in reality can be placed to produce understanding. Seeing fish recalls fish. Seeing fowl recalls fowl. Thus, when something presents itself in reality for which we have no recall, arguments develop as to its classification. Until there is a classification, a level of recall at which to place something, there will be no feeling of knowing when we see that something in reality.

The process of empirical science is one in which these arguments are resolved, with a fin ignored here, a feather created there, or an entirely new classification established.

The problem with the process is that we are making up the categories in which to fit what we see.

We are just naming things.

We are understanding nothing!

However, because we are able to settle disputes about the nature of objects in reality by agreeing as to the classification of the objects, and because the agreement of recall with reality produces a feeling of comprehension, empirical science, misapprehending the search for comprehension about the nature of physical reality, the hidden motion of things, believes that it can also solve disputes as to the nature of the forces that cause movement by simple agreement.

Because we can always make up categories in which to classify real objects in reality, understanding the things we see by the names we create for them, we think we can make up words and apply them to the things for which we have no explanation, falling objects, rotating and orbiting planets, light breaking down in a prism and by monkey saying the words when we monkey see the thing happening, we will know what is happening.

Empirical science, with its process of turning concepts into facts by the application of words, makes us believe that we know everything there is to know about something when we know absolutely nothing about it!

Of course, naming things might be all the knowledge we need to eat, but when it comes to figuring out why objects fall, why the planet rotates on it axis, or what light is, making up categories and naming things that fit into the categories does not produce useful knowledge. Creating categories for occurrences in physical reality, falling objects, rotating planets, expanding flows of light, is not the same as categorizing objects.

Creating mass/gravity as a category for falling objects so that whenever we are confronted by falling objects we recall mass/gravity doesn't tell us anything about falling objects. But because we can put a name to falling objects, we have the feeling of knowing and we delude ourselves into thinking we know. When we think we know something, we don't look for answers because we have no questions. Creating a category for rotating planets, angular momentum, doesn't tell us anything about why the planet is rotating, it just uses words to name what is occurring. Objects in physical reality are there to be named. Occurrences in physical reality have to be described. Angular momentum doesn't describe anything.

Making up words to classify light doesn't tell us anything about how light operates. In fact, by its process of classifying and categorizing, empirical science separates heat from light, categorizes it with effects on a planet's surface, measures it by the amount that strikes a given area of physical reality, basis its brightness on a process of grading an arbitrary magnitude, and misses totally what is happening to it dynamically. The empirical process, satisfied with the mindless knowing involved in classifying and categorizing, is incapable of seeing that light expands over the surface of an expanding sphere, and as it does so, diminishes in the same manner we measure the force that causes objects to fall, inversely with the square of its distance form the source.

When we mindlessly create categories, and then delude ourselves into thinking we comprehend by placing names to things in the category, we miss the measurable correlation between the two most important physical phenomena in our realty, the fact that emission fields diminish in the precise mathematical way that objects accelerate in those fields.

As a result of deluding ourselves into believing that naming things to categories produces comprehension, our scientific journals and magazines, our books and newspapers are filled with pictures, drawings, photographs and other electronic representations, many given false colors and delineations that don't exist, representations that defy description. Whether they be blobs of light or radiation from outer space, shadowy representations from the body's innermost recesses or swirls and vortices from atom smashers, we have to read the nicely typed words printed below the picture to make sense of what we see.

We see "Birth of Stars in the Crab Nebula", or "Gene that Produces Aggression" or "Electron Binding Atoms" and we comprehend.

We have no way of taking a picture of a star being born, we have no way to see a gene causing aggression, and we certainly have no way of seeing an electron doing anything, but because we sign a picture with a shibboleth, we think we comprehend. In empirical science, the only way we can comprehend is to pass this scientific Rorschach test. If we can't repeat the common words, if we see nothing, or worse, if we see reality, we're locked up as crazy.

Putting names to what we see doesn't produce comprehension of the process itself, explain what is actively happening in physical reality. Saying white light goes in one side of the prism, and motley light comes out the other side does not satisfy the need for comprehension. We want something more, so we say that the white light that went in must contain all of the colors that came out, else where did the colors come from? This is as empirical as we can get. Who can disagree with the conclusion?

When we all agree that something that can't be measured has to be true, we have established a consensus about reality, created a scientific fact.

From the point that we accept the conclusion as comprehension for what is happening when white light produces colored light in a prism and place it in our recall, we cease to see reality and begin to see made-up reality, because from that point on, our recall is stored at the same level of excess affinity propensity as light is stored so that every time we see light, we recall that light is made up of all colors, and we recall nothing else. We have no ability to recall something other than what is recalled by the current generated by what we see, so we have no reason to make the effort to recall anything else because we don't have any desire to recall something that would cause us to cease to comprehend and recalling that light is made up of all colors allows us to comprehend what light is.

Modern science, which I distinguish from empirical science in that empirical science attempted to base its conclusions on fact whereas modern science has no factual basis whatsoever, modern science is filled with these little bon mots, the Earth rotates on its axis because angular momentum is conserved in closed systems, objects move in straight lines unless a force acts upon them to change that motion, planetary rotation is the result of the solar system condensing out of a swirling mass of gas, gravity is proportional to and a property of mass, all of which satisfy the need to comprehend something that isn't measurable and none of which contains any meaning, any explanation of what is actually going on in physical reality.

Facts are not inherently meaningful, and explanations for facts can never themselves be facts because explanations are what our mind devises and places in our recall in order that our recall can recall something to explain the facts that have no inherent meaning.

We can devise tools to measure facts, the hard edges of reality, but we cannot understand those facts without the concepts that allow our minds to construct pictures that show how those facts interact to form the reality that we see and can measure.

Concepts, then, are only as good as the facts that they explain, and any process we devise does not need to deal with facts, which are easily measurable, it needs to deal with the concepts we devise, which can never be facts, but which can only be used to comprehend the facts.

The Copernican conclusion, that the Earth travels around the sun, took an extreme amount of ingenuity to become a physically measured fact. However, just because the pre-Copernican misconception turned out to be provable as a fact doesn't mean that all of the other things we want to know can also be proven to be fact.

We will never know, with factual certitude, what Bacon called the secret motion of things. We can only produce concepts that approximate what we can measure about them. To believe we know with factual certitude what causes objects to fall, or what causes planets to move, or how light is structured, starts us down the road to delusion.

When we mistake a concept for a fact, white light contains all colors, and put it into our recall so that every time we see white light we recall the concept, light contains all colors, we cripple ourselves when we go on to examine what is going on in the reality that the light strikes. When we see light striking the leaves of a tree and the light come back green, while the same light is striking the trunk and that light is coming back brown, we don't ask what is it about light that allows it to strike one thing and come back green and strike another thing and come back brown, we ask what it is about the green thing and the brown thing that allows it to reflect back green and brown light when light that contains all the colors of the rainbow strikes them.

When we know that light contains all the colors that can exist, and we are seeing only green and brown light reflected, we don't ask what change the light has undergone, we ask what happened to all the other colors!

And, with the fact that matter absorbs colors firmly in recall, a conclusion dictated by our assumption that light contains all colors, when we become capable of reducing light to its smallest flows by using affinity propensities to direct electron flows, and we start to see those flows bounce off of the matter that light reflects off of, we recall absorption and create electrons in the atoms we have also made up and stored in recall, jumping around absorbing light.

And, if the electrons can absorb light by jumping around, jumping around the other way will give light off!

We did the same thing when we made up some nice little statement about the velocity of light being inversely proportional to its refractive index to describe why blue light bent in the prism at a greater angle than green, and green, yellow and yellow, red. When it came time to put the made-up notion in our recall that light was the same as water waves, every time we attempted to measure the distance between the crests of the newly created light waves, our recall dictated the outcome, putting blue at the shortest wavelength because it refracted the most and red at the longest wavelength because it refracted the least.

There was absolutely no restraint in physical reality from our doing so because we had made up everything from the get-go, and the only thing our fantasies had to avoid disagreeing with was that white light came in one side of the prism and left the other as different colors.

And even though blue is clearly the lowest frequency of visible light, it's additive to both green and red to increase their frequencies respectively, and additive to green and red, yellow, to produce the uncolored frequency of white light, it is the only color we see once we have been frequency adapted to yellow and its luminescence goes from yellow as the brightest to blue as the dimmest, when it came time to show how far these made-up electrons jumped in the made-up atom to produce and absorb light, they did their most energetic jumping to produce and absorb blue light, and their least energetic jumping to produce red.

We do not see reality with our eyes, we see it with our recall, and if we don't have any way to control the stuff we put into our recall, other than to say, well, the idea predicted a fact that turned out to be a fact, so the idea must be a fact too, then we are going to be in sad shape.

But, then, we are already in sad shape.

Just how sad a shape are we in?

The only thing that approximates reality about the concept of evolution as developed in Darwinism is the concept of evolution itself. The idea that one species evolves into another species is as absurd as the idea that white light is made up of all the possible colors that the eye can see, or more accurately, all frequencies.

Atoms and molecules of atoms form around electrical flows in the environment to satisfy the excess affinity propensities of both the flows and the atoms and molecules. This is the animate matter we categorize as life. It forms to match the conditions in the environment that gave rise to its formation.

If this were all there was to it, then life would simply form around existing flows, and exist as long as those flows existed, ceasing to exist when the flows ceased to exist.

However, this isn't all there is to it because the flows around which the atoms and molecules of atoms are forming exist on a planet whose movement is dynamic with respect to the star it orbits. It rotates on its axis, which provides a continuous change in the potential differences of the matter randomly distributed on its surface. These changes continually alter the electric flows around which the atoms and molecules of atoms are forming.

More important, the Earth is cooling. The Earth's surface environment is undergoing continuous change, a steady reduction in temperature as a result of the slow thickening of the crust as the combusting sphere of matter beneath its surface recedes, radiating its heat out into space.

As described in *The Cooling Continuum*, *The Rise and Fall of Species on Earth*, the second volume of The Copernican Series, the environment in which life is coming into existence on Earth has been and is continually undergoing change, giving rise to new forms of life. Because any successful process that organizes atoms and molecules of atoms around electrical flows in the environment is repetitive, the process of organizing those atoms and molecules of atoms can be encoded as genetic information. This genetic information is not information about species, but rather information about characteristics that have been successful at one point or another in the cooling continuum that is the evolving environment. Whether these characteristics deal with locomotion, the organization of functional cilia or paddles, or sensing the environment, the production of minute compounds of atoms which sense the heat in the environment by field replacement, or about boundary matter, the construction of tissue that separates an organism's functions from the environment, these characteristics become genetic material that can be passed on through the chain of evolution.

It is not species, nor is it categories, classes or any other words we dream up to put what we currently see in neat compartments, that evolve, it is characteristics that develop and that are passed on to subsequent generations encoded as genetic matter that result in evolution.

Evolution, therefore, is not about evolving species, but rather about evolving complexity.

And the purpose of complexity is to extend the range of survivability of species within the environment!

As the range of genetic characteristics upon which animate matter can draw to adapt to the continually changing circumstances in the environment increases, the organisms that can form to take advantage of those conditions become better able to develop characteristics, or to conglomerate existing characteristics into new characteristics to extend their range of survivability within the environment.

The force driving evolution, however, is one way, into a cooling environment, and thus evolution is a one way trip, from simple to complex, from dependency on the environment toward extended survivability independent of the environment.

The Darwinian insistence that evolution has no

purpose is born of the failure to understand the environmental changes in which evolution is occurring and as a result, it focuses on survival as its ultimate operative mechanism. However, it is not that simple. It is not survival that drives evolution, but rather the effort to extend survivability.

Extending survivability within an environment means conglomerating characteristics into organisms that are increasingly capable of existing independent of the environment.

We can see this evolution of characteristics clearly by simply observing what has in reality happened on the Earth since the first animate matter formed. We have the continuous spectrum around us at all times.

Animate matter can be so simple that it is totally at the mercy of its environment. We can place a drop of water under a microscope and see it teaming with animate matter, life. We can boil the same drop of water and see the animate matter come to an end. Animate matter that is totally dependent on its environment will not survive changes in that environment.

The statement is so obvious that it needs repeating.

Animate matter that is totally dependent on its environment will not survive changes in that environment.

That is why we have evidence of universal extinctions in the fossil record as the cooling process of the planet passed the range of survivability of one generation of animate matter and moved into the range of survivability of another generation of animate matter.

These ranges of survivability are what allow animate matter to develop characteristics that extend its range of survivability beyond the normal range of the environment. Trees, for instance, have broad ranges of survivability, far broader than lichen. However, nonambulatory animate matter is still at the mercy of what happens in the environment. If fires rage, trees will die.

Ambulatory animate matter, with legs and wings and fins, can move within the environment, however, and thus extends its range of survivability within that environment. Animals can flee before the raging temperatures of the forest fire.

And, to a certain extent, ambulatory animate matter can alter the environment in order to extend its range of survivability within that environment. Beavers can build dams, squirrels can store nuts, birds can build nests.

The characteristics that allow animate matter to alter the environment, however, require the development of a series of complex characteristics that can form pictures of how reality is, and then store those pictures for future recall. At which point animate matter is merely evolving to take advantage of the environment, a clam burrowing into the sand, and animate matter is purposely changing the environment outside of merely expressing genetic characteristics is like arguing how smart dogs are.

However, developing a complex of characteristics that

allows the formation of a picture of reality, that allows the storing of that picture, and allows the subsequent recall of that picture is of immense value in extending an organism's range of survivability within the environment because it allows the organism to form pictures of the environment when the environment is not there. Once an organism can form a picture of the environment without the environment present, it can change that picture by altering the current level of its individual elements so as to recall different elements.

An organism that can form a picture of reality that doesn't exist in reality can extend its range of survivability by forming pictures of the environment that would be more compatible to its survival, and then altering the environment to match those pictures. Instead of merely fleeing the forest fire, sentient animate matter would be able to find ways to put the fire out.

It is not uncommon to see forms of animate matter other than man develop tools to alter the environment in order to extend its respective ranges of survivability within that environment, but man has developed the complex of characteristics that accomplishes this goal to the greatest degree. We can form pictures of reality, alter the current levels that are recalling individual elements of those pictures to the smallest or greatest degree, producing pictures of reality that might or might not be capable of existing in physical reality.

Determining what might or might not work in reality is the basic trial and error process of the scientific method and we use this basic scientific method to produce the technology that allows us to extend our range of survivability independent of the environment.

We embody the entire range of evolution. We are all of the evolved characteristics that will ever develop on this planet. We incorporate the complexity that allows us to alter the environment in order to extend our range of survivability within that environment and thus to extend the range of survivability of all of the characteristics that evolved on this planet.

However, extending our range of survivability depends on how successful we are in developing a technology that can alter the environment in a way that will extend our range of survivability within that environment. Developing a technology that will allow us to alter the environment in a way that will extend our range of survivability within that environment requires us to have an accurate picture of that environment.

If we do not have in our recall an accurate picture of the environment, then any attempt to alter the individual elements of that picture will be futile because, not having an accurate picture to start with, we will be attempting to alter a picture of a reality that doesn't, in reality, exist.

The only way that we can understand external reality is through the use of our recall. We see what we recall, and as long as recall matches reality, we can move on our way through that reality, the basic purpose for which the complex system of recall evolved.

However, we cannot see what we cannot recall, and when we put stuff in our recall that co-opts reality, that provides recall for a reality that doesn't exist, then we will never be able to use our recall to understand the reality that actually exists.

When the purpose of our recall is to move ourselves through physical reality, then we have a self-correcting process. We bump into reality and reality intrudes. We have to alter our picture of reality if we wish to continue to move through that reality.

However, there is much about reality we don't bump into.

Until we started sending rockets into space, it made little difference whether the Earth moved around the sun, or whether the sun moved around the Earth.

If we form inaccurate pictures of reality, and we never bump into the reality that those pictures represent, then we can just go our merry way, ignorant of reality.

We don't bump into some of the most prominent aspects about our reality.

Not knowing why objects fall does not prevent us from moving through reality. If we knew what makes objects fall, of course, we might be able to devise a method to better move through the environment, but when we think we know what makes objects fall, and that makes us know that we can't devise a method to better move through the environment, we will simply not attempt to produce technology that will allow us to utilize the force that causes objects to drop.

Not knowing what makes the Earth rotate on its axis doesn't stop us from feeding our faces, so we never attempt to form accurate pictures of what is going on as the Earth turns on its axis. We have no idea what makes planets orbit the sun, but not knowing what makes planets orbit the sun doesn't keep us from moving our bowels on a regular basis, although having a physical description of the forces that cause rotation and orbiting might well allow us to form pictures that would lead to the production of technology that would allow us to extend our range of survivability.

Having no idea what light is, how it is produced, or how it acts on matter does not keep us from enjoying the sunset, although knowing about light might well allow us to extend our range of survivability beyond the final sunset.

When we make up things about physical reality that have no basis in physical reality, when we make up half-assed explanations for physical phenomena that make us believe that we understand what is going on with respect to those physical phenomena, we limit the picture of reality that we see with our recall and therefore limit our ability to change that picture in order to change reality to accommodate our existence.

When we make up mindless explanations like gravity is a property of matter, angular momentum is conserved in closed systems, planets condense from swirling masses of gas, light travels in waves, white light contains all colors, the wind blows because of the Coriolis effect, and the deadliest of all, we can prove concepts to be facts by finding facts the concepts predict, we cripple our ability to extend our range of survivability in reality by limiting our recall to made-up stuff.

When we fill our recall with bunk, the picture of reality we form is bunk, and the technology we create in furtherance of that picture is distorted at best, if indeed it exists, because it is more likely simply not to exist. We use combustion to push rockets against the force that makes objects fall instead of analyzing the force that makes objects fall in the first place and then creating a technology that utilizes that force.

By creating recall that doesn't exist in reality, we blind ourselves to reality.

We simply don't see the reality in which we exist!

So, who cares?

If we don't bump into obstacles in physical reality, so much the better. We can just keep moving through physical reality like we have done for billions of years and like we can continue to do for billions of more years.

The idea that life has been around for billions of years and will continue to be around for more billions of years into the future is the result of putting more bunk in our recall, the concept that evolution involves morphing from one species to the next. Fish don't become birds, or vice versa, in a day, a year or even a million years, we say. No, for species to evolve from one species to the next takes billions of years, and for proof, we only have to consult our recall, where the fact obtains common agreement.

However, anyone dealing with evolution today can't fail to note the rapidity with which species evolve to accommodate the environment in which they exist, and the only thing that keeps species within their proper bounds is the flexibility of our definitions as to what is categorized as what. As the title of a chapter in *The Cooling Continuum* so aptly puts it: "All At Once, Not All From One." When conditions exist that allow simple animate matter to exist, that animate matter comes into existence through the spontaneous formation of atoms and molecules of atoms around electrical flows in the environment, not by individual conjugations through the eons.

If we want to know the history and future of the Earth, we have but to look at the reality around us, the planet Venus, which approximates a young Earth, and the moon which approximates a dead Earth. We are somewhere in between the two, I suspect closer to the moon than to Venus.

The spectrum of the evolution of life is found between the congealing surface of Venus and the dusty crust of the moon, and the time scale between the two is not in the range of a billion years, nor even in the range of a million years, Lord Kelvin's estimates eschewing biblical spans that would bring him collegiate criticism rather than favoring physical accuracy.

The time span between Venus and the moon is probably more in the range of fifty to a hundred thousand years!

We can already measure the oceans rising, and in true empirical form, where we see with a recall that is opposite reality, attribute it to global warming and melting ice caps rather than the receding sphere of combustion beneath the surface of the Earth. As the Earth cools, the strength of its emissions are reduced as they pass through the thickening crust. As the strength of the emissions decrease, the by-product of the mechanism that regulates their uniform expansion, the by-product which produces the attractive force that holds matter to the surface of the Earth, also decreases. With the force of attraction weakening, the water in the oceans is naturally becoming less compressed and the surface of the oceans is slowing rising.

The planet is not heating up, it is cooling off. The ice caps are not getting smaller, they are getting larger.

An easier way to plot the Earth's rate of cooling is to purchase a book listing world records and plot the ever improving weight related records.

Not only do we misapprehend the nature of the environment in which we exist, we misapprehend the environment in which we have to extend our range of survivability.

We are accustomed to using our minds to construct

pictures of the physical reality in which we exist, altering that physical reality, building skyscrapers, tunnels, roadways, producing automobiles and airplanes, engineering information systems and communications infrastructures, and we are totally unaware that the environment we are spending our time altering is not the environment in which we have to extend our range of survivability.

We may be accustomed to sunlit days and balmy nights, but the environment in which we exist and in which we have to extend our range of survivability, if we are going to carry the unique life that developed on the face of this planet into the future, is the endless void of nothingness that surrounds the Earth, cold, empty space.

If we are going to survive, if we are going to extend our range of survivability into the future, then we are going to have to come to terms with what we have put into our recall to explain physical reality, and how we came to put it there.

We are going to have to come to terms with what we are and what we can do with ourselves.

We are going to have to see ourselves for what we are, and we are going to have to see how what we are relates to the physical reality that produced us, a physical reality which, unless we get off our mental duffs, will reclaim us.

To see or not to see is not the question, it is the choice, and it is a choice between comprehension and survival or ignorance and extinction.

We literally have to see the light if we wish to continue to exist.

So who cares?

Only those interested in the continued existence of the unique life that evolved on the face of this planet.

Only those interested in extending our range of survivability so that we may take our proper place in the universe.

We are on a clock and that clock is ticking every instant of our existence, the ticking of the Earth's emissions, radiating out into space at the rate of trillions of calories a second, second after second, moving us closer each second to the end of our own range of survivability on the surface of this planet.